Nelson Maths

2

W0090530

This book belongs to:

_____

Workbook

Karen Morrison
Lisa Greenstein

OXFORD
UNIVERSITY PRESS

OXFORD
UNIVERSITY PRESS

Great Clarendon Street, Oxford, OX2 6DP, United Kingdom

Oxford University Press is a department of the University of Oxford.

It furthers the University's objective of excellence in research, scholarship, and education by publishing worldwide. Oxford is a registered trade mark of Oxford University Press in the UK and in certain other countries.

British Library Cataloguing in Publication Data

Data available

ISBN: 978-1-382-01026-9

1 3 5 7 9 10 8 6 4 2

Paper used in the production of this book is a natural, recyclable product made from wood grown in sustainable forests. The manufacturing process conforms to the environmental regulations of the country of origin.

Printed in Great Britain by Bell and Bain Ltd, Glasgow

Acknowledgements

The publisher and authors would like to thank the following for permission to use photographs and other copyright material:

Cover: Matthieu Nivesse.

Artwork by Liliana Perez, Q2A Media. Pantek Media, and OKS Prepress.

Every effort has been made to contact copyright holders of material reproduced in this book. Any omissions will be rectified in subsequent printings if notice is given to the publisher.

Contents

Unit 1 Think maths   **4**
Mistakes and your brain 4

Unit 2 Working with numbers   **5**
Count and show numbers........................ 5
Count to 50 6
Count and write 7
Missing numbers 8
Number lines 9
Estimate and count 10

Unit 3 Place value   **11**
Tens and ones 11
Make and break numbers 12
Place-value cards 13
Use the = sign 14
Compare numbers 15
Order numbers 16

Unit 4 2D and 3D shapes   **17**
3D shapes 17
Which shape is it? 18
Properties of 3D shapes....................... 19
Sides and vertices 20

Unit 5 Patterns and sequences   **21**
Shape patterns 21
Pattern rules..................................... 22
Number patterns 23
Skip-counting patterns 24

Unit 6 Add and subtract   **25**
Bar models.. 25
Add tens... 26
Subtract tens..................................... 27
Count on in tens and ones 28
Use a 100 chart to add and subtract 29
Make 100.. 30
Use number facts to add 31
Add four or five numbers 32

Unit 7 Length   **33**
Measure lengths................................. 33
Estimate lengths 34
Measure in centimetres 35
More centimetres 36

Unit 8 Mass   **37**
Heavier or lighter 37
Kilograms.. 38
Measuring kilograms 39
Animal masses 40

Unit 9 Lists and tables   **41**
Sort fruits and vegetables.................... 41
Use tally tables 42
Sort numbers 43
Sort shapes....................................... 44

Unit 10 Show data   **45**
Pictograms.. 45
Draw a block diagram 46

Read a block diagram 47

Unit 11 Multiply   **48**
Group in different ways 48
Add groups 49
Rows of 2.. 50
Arrays.. 51
Multiply by 2 52
Odd and even..................................... 53
Multiply by 5 or 10 54
Make groups 55

Unit 12 Divide   **56**
Equal shares 56
Divide equally 57
The division sign 58
Multiplication and division facts............ 59

Unit 13 Fractions   **60**
Make equal parts................................ 60
Write fractions 61
Half of a group 62
Quarter of a group 63
Halves and quarters 64
Fraction problems............................... 65

Unit 14 Time   **66**
Seconds, minutes and hours.................. 66
Months of the year 67
Compare times 68

Unit 15 Possible outcomes   **69**
Possible outcomes............................... 69
Regular and random patterns 70

Unit 16 Symmetry   **71**
Line symmetry.................................... 71
Reflections 72
Symmetrical patterns 73

Unit 17 Capacity and temperature   **74**
The litre 74
Estimate and measure 75
Temperature...................................... 76

Unit 18 More about time   **77**
Tell the time...................................... 77
Quarter hours.................................... 78
What is the time?............................... 79
More times.. 80
Write digital times 81

Unit 19 Position and movement   **82**
Moving in a straight line....................... 82
Turning left and right 83
How many turns?................................. 84
Quarter turns 85

Unit 20 Money   **86**
Our currency...................................... 86
How much change?............................... 87
How much is it worth?.......................... 88

Mistakes and your brain

THINK MATHS prepares you to think mathematically, value mistakes, and learn maths with a growth mindset.

1 On the picture, show what happens to your brain when you make mistakes. You can add colours, more pictures and words.

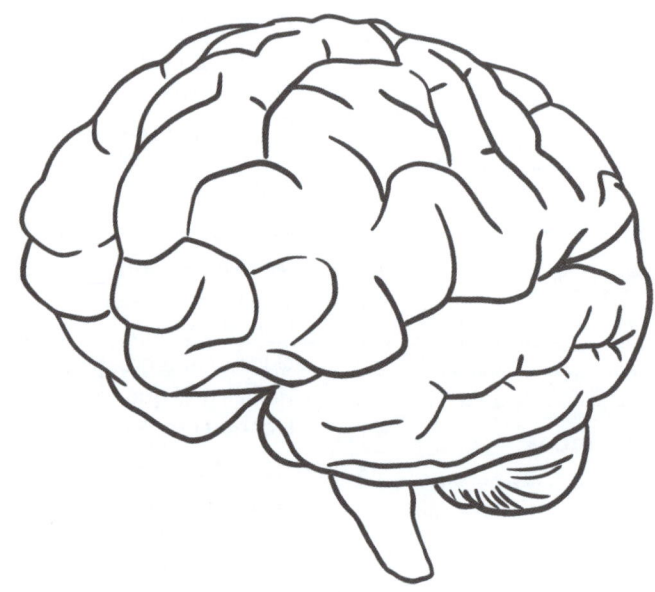

2 Complete the sentences in your own words.

a Everyone can do maths because _____

_____

_____.

b Mistakes are important because _____

_____

_____.

➤ *Pupil Book page 6*

Working with numbers

Count and show numbers

1 Find pairs that show the same number.

Colour the matching groups the same colour.

A ○ ○ ○
  ○ ○

B ○ ○
  ○ ○

C ○ ○ ○ ○
  ○ ○ ○

D ○
  ○ ○ ○
  ○

E ○
  ○
  ○ ○

F ○
  ○ ○
  ○ ○
  ○ ○

G ○ ○ ○
  ○ ○ ○

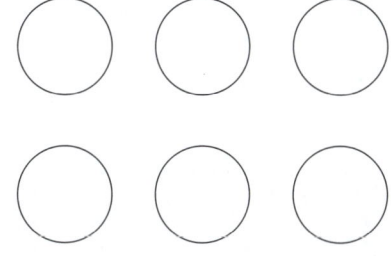

H ○ ○
  ○
  ○ ○

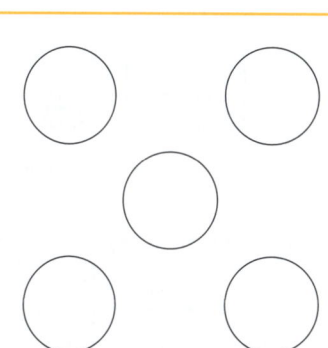

➡ *Pupil Book page 8*

Count to 50

1 Write over the grey numbers.

1	2	3	4	5	6	7	8	9	10
11	12	13	14	15	16	17	18	19	20
21	22	23	24	25	26	27	28	29	30
31	32	33	34	35	36	37	38	39	40
41	42	43	44	45	46	47	48	49	50

2 **a** Say all the number names in the chart out loud.

    b Count back from 50 to 1.

---

3 Fill in the missing numbers. Use the 50 chart to help you.

a

b

c

d
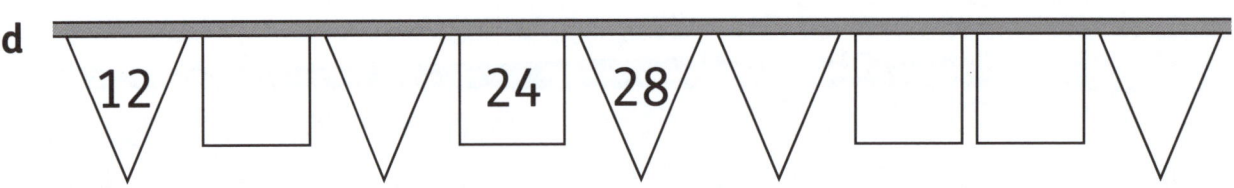

➡ *Pupil Book page 10*

Count and write

1 Make groups of 10.

Write the number.

Write the number name.

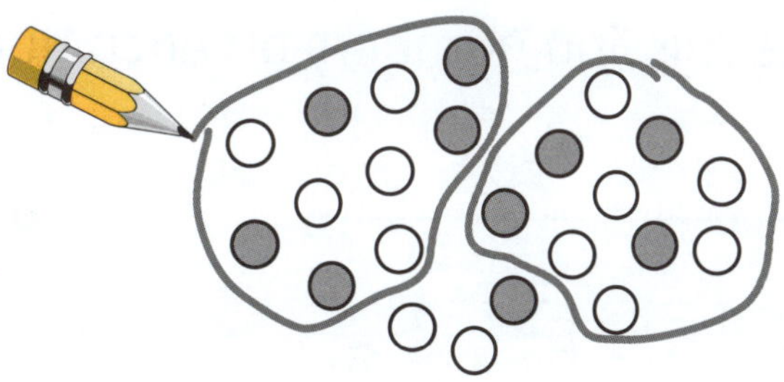

23

twenty-three

a

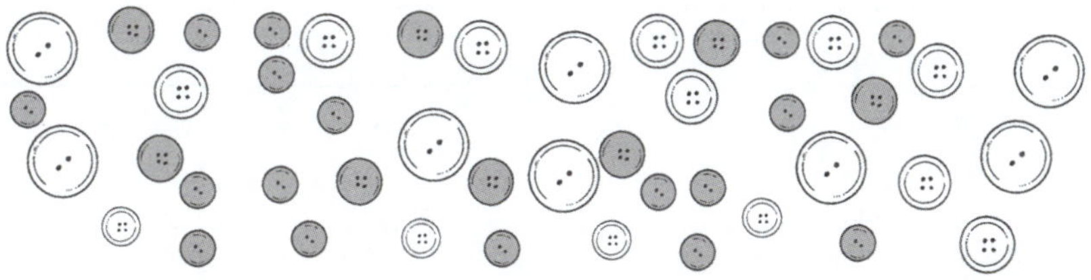

b

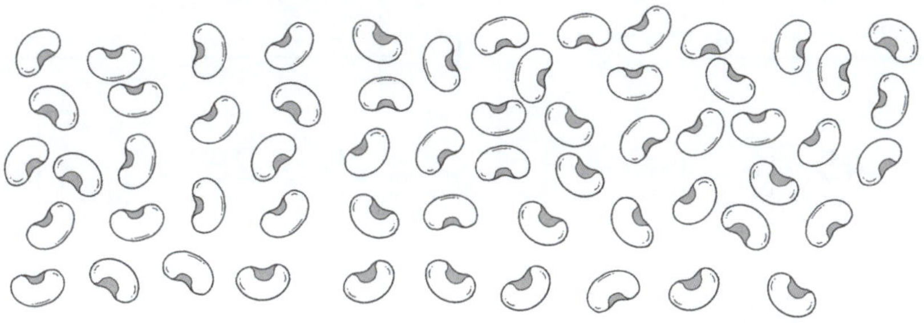

c

d

e

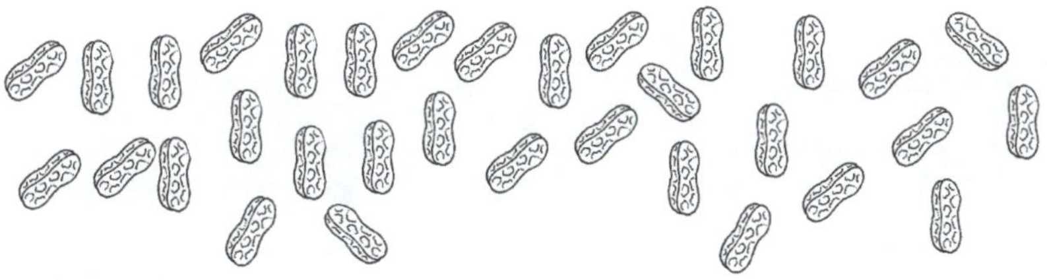

➡ *Pupil Book page 11*

Missing numbers

1 Fill in the missing numbers on each number line.

a 27 ☐ ☐ 30

b 32 ☐ 30 ☐

c 37 ☐ ☐ 40

d 43 ☐ ☐ 46

e 47 ☐ 45 ☐

f 29 28 ☐ ☐

g 50 ☐ ☐ 47

h 39 38 ☐ ☐

i 14 16 ☐ 20

j 21 ☐ 25 27

2 I am thinking of a number.

It is between 30 and 40.

It is less than 38.

It is greater than 36.

What number is it? ☐

➡ *Pupil Book page 12*

Number lines

1 Choose a number between:

10 and 20 | 14   20 and 30 | 23

Show your numbers on this number line.

```
0              10      14        20    23        30
|---------------|-------|---------|-----|---------|
```

a Choose a number between:

20 and 30 |    40 and 50 |

Show your numbers on this number line.

```
20              30              40              50
|---------------|---------------|---------------|
```

b Choose a number between:

30 and 40 |    40 and 50 |

Show your numbers on this number line.

```
20              30              40              50
|---------------|---------------|---------------|
```

c Choose a number between:

60 and 70 |    80 and 90 |

Show your numbers on this number line.

```
60              70              80              90
|---------------|---------------|---------------|
```

➡ *Pupil Book page 12*

Estimate and count

1 Estimate how many balls.

Circle your estimate.

Count and write the total.

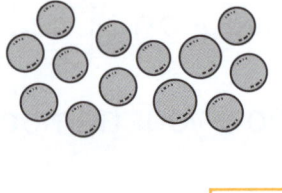

13

---

a 10, 20, 50 or 100

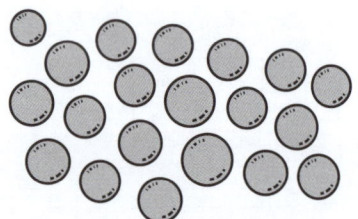

b 10, 20, 50 or 100

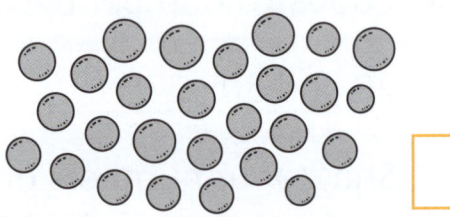

---

c 10, 20, 50 or 100

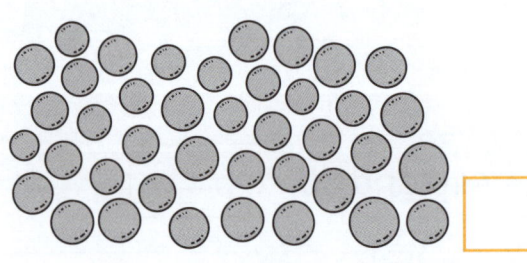

d 10, 20, 50 or 100

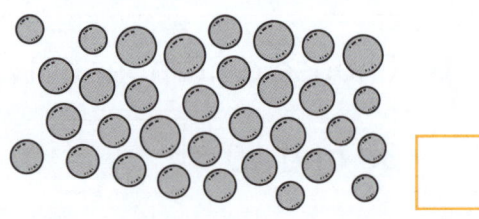

---

e 10, 20, 50 or 100

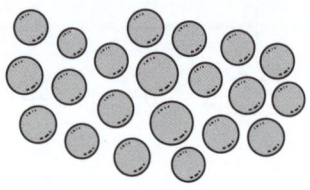

f 10, 20, 50 or 100

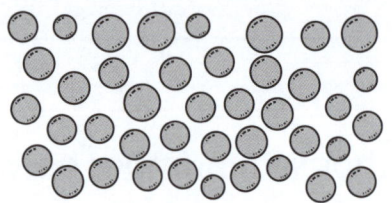

---

g 10, 20, 50 or 100

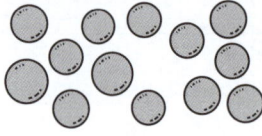

h 10, 20, 50 or 100

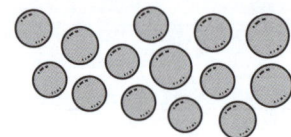

➡ *Pupil Book page 13*

Place value

Tens and ones

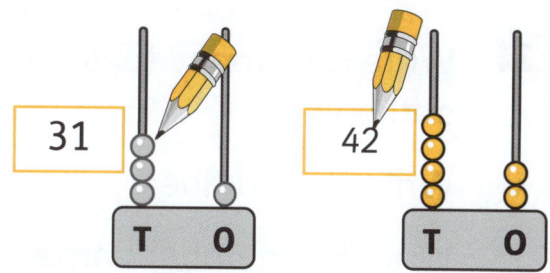

1 Draw the beads or write the number.

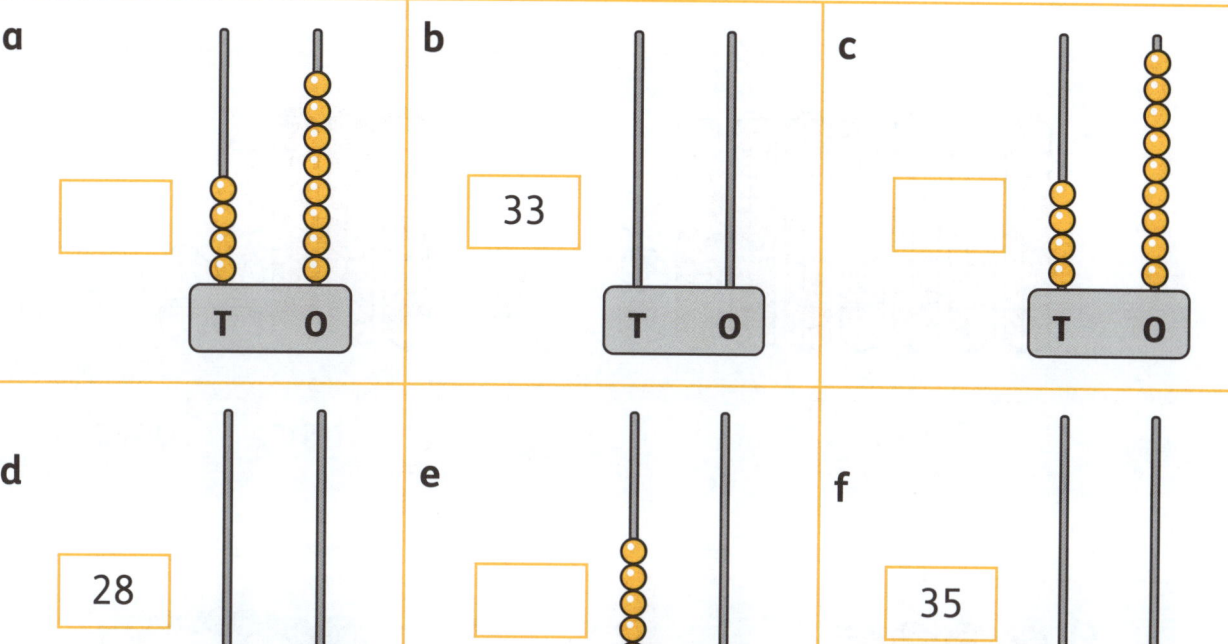

a	b 33	c
d 28	e	f 35

2 How many tens and ones? The first one has been done for you.

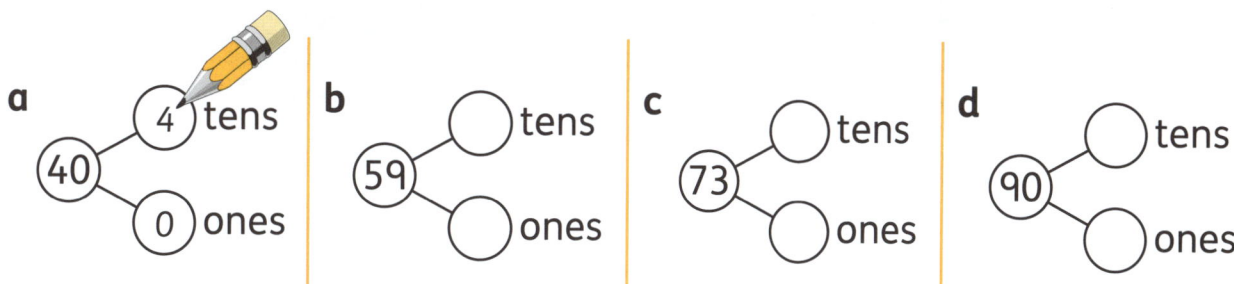

a 40 → 4 tens, 0 ones
b 59 → tens, ones
c 73 → tens, ones
d 90 → tens, ones

3 Maria has 45 marbles.

a How many groups of 10 does she have? ☐

b Jasmine gives Maria 8 more marbles.

How many groups of 10 does Maria have now? ☐

➡ *Pupil Book page 14*

Make and break numbers

1 Write how many tens and ones.

Write the number.

Write the number name.

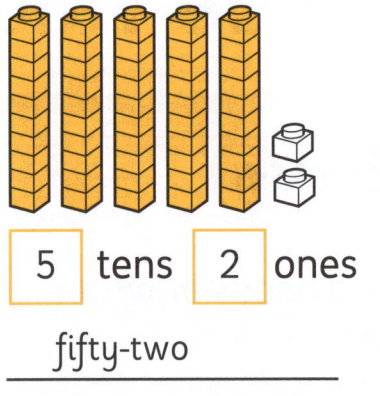

5 tens 2 ones

fifty-two

52

a

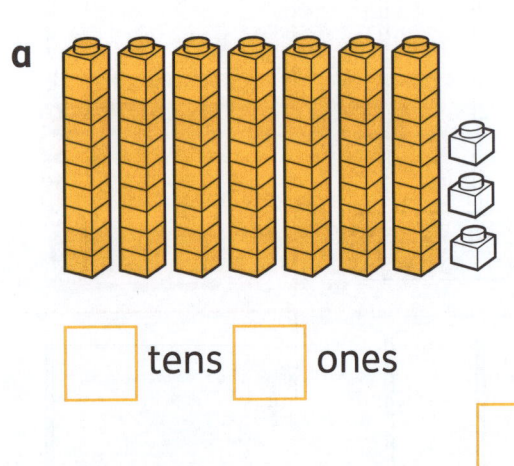

☐ tens ☐ ones

_____ ☐

b

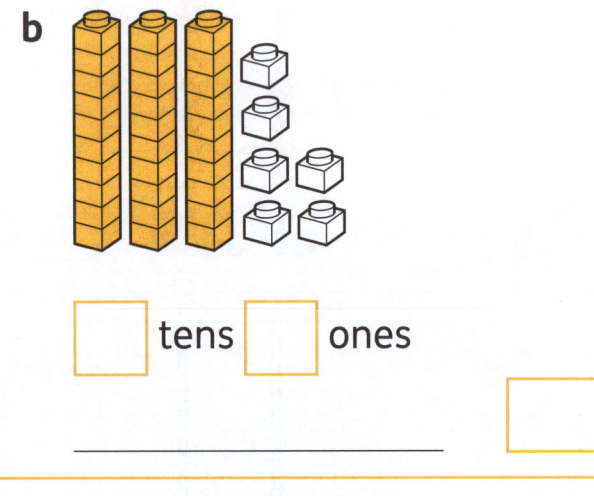

☐ tens ☐ ones

_____ ☐

c

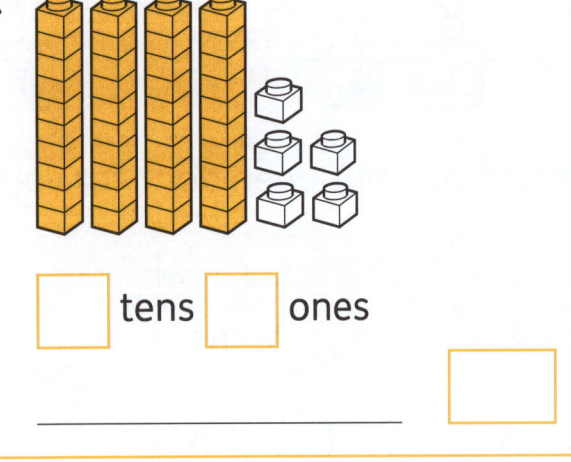

☐ tens ☐ ones

_____ ☐

d

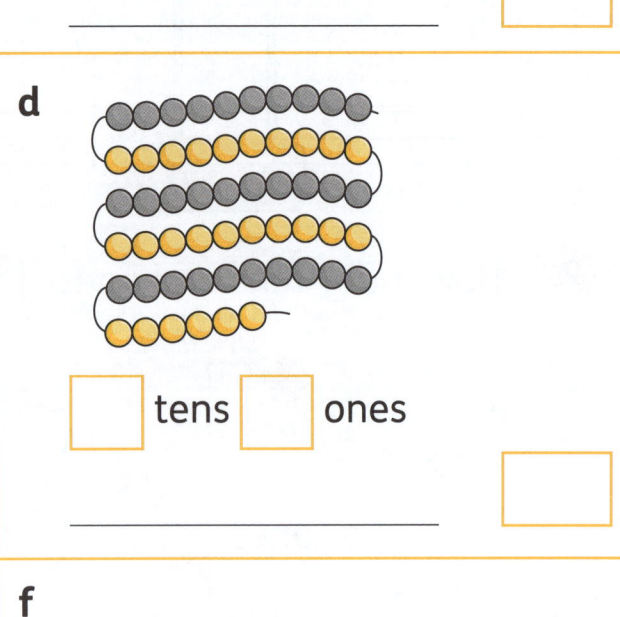

☐ tens ☐ ones

_____ ☐

e

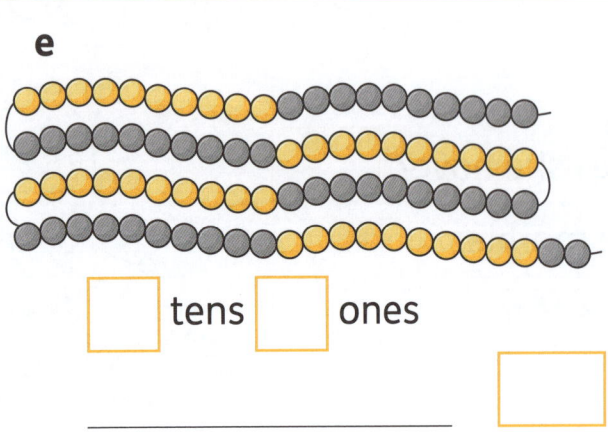

☐ tens ☐ ones

_____ ☐

f

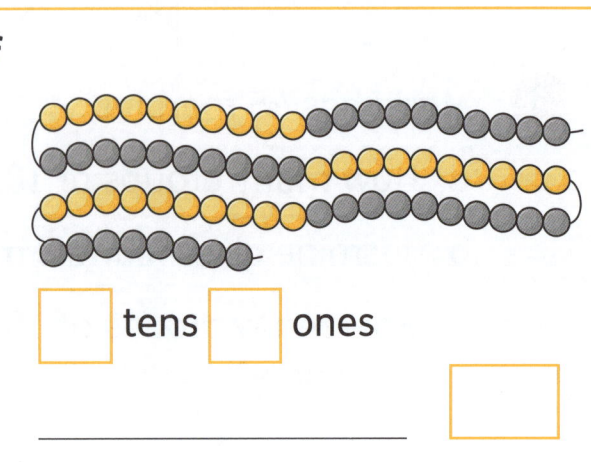

☐ tens ☐ ones

_____ ☐

➡ *Pupil Book page 15*

Place-value cards

You can use place-value cards to make numbers with tens and ones.

Start with the place-value card that has zero in the ones place.

Slide the ones card over the zero to make the new number.

1 Write the addition.

The first one has been done for you.

a
| 3 | 0 |

and

| 7 |

30 + 7 = 37

b
| 4 | 0 |

and

| 5 |

_____

c
| 8 | 0 |

and

| 3 |

_____

d
| 1 |

and

| 7 | 0 |

_____

2 Say the number. Complete the tens and ones table.

a
| 7 | 2 |

b
| 3 | 8 |

c
| 8 | 8 |

d
| 8 | 0 |

e
| 6 | 9 |

f
| 4 | 9 |

	Tens	Ones
a		
b		
c		
d		
e		
f		

➡ *Pupil Book page 16*

Use the = sign

1 Complete each picture or addition to make it equal.

a

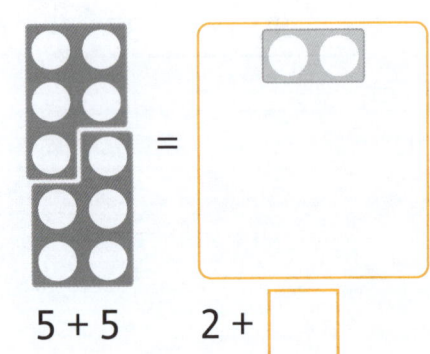

= ☐ tens

b

+ ☐ = 1 ten 5 ones

c

5 + 5   2 + ☐

d

6 + 6   ☐ ten ☐ ones

2 Now make your own pictures or additions to show equal amounts.

a

b

➡ *Pupil Book page 17*

Compare numbers

Remember, < means less than and > means greater than.

1 Write < or > between the pictures.

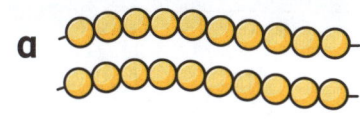 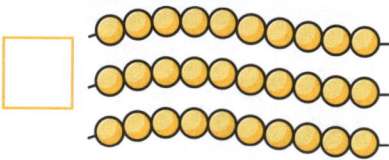

a   2 tens   3 tens

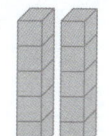 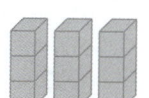

b

2 Write < or > between the numbers.

a 27 ☐ 31   b 68 ☐ 86   c 53 ☐ 55

d 37 ☐ 27   e 83 ☐ 78   f 99 ☐ 100

3 Write <, > or = between the number sentences.

a 10 + 30 ☐ 50 − 10   b 30 − 10 ☐ 20 + 10

c 50 + 10 ☐ 80 − 10   d 40 − 10 ☐ 40 + 10

4 Put a ✓ if it is correct. Put a ✗ if it is not correct.

a 43 > 41 ☐   b 23 > 32 ☐   c 37 = 30 + 7 ☐

d 35 < 45 ☐   e 3 = 30 ☐   f 14 < 41 ☐

5 Circle the number that does not fit in each box.

a | 30 40 50 55 60 70 | b | 27 26 31 25 24 23 |

c | 42 44 46 48 49 50 | d | 100 98 96 95 94 92 |

➡ *Pupil Book page 18*

Order numbers

1 Write each set of numbers in order from smallest to greatest.

a 19   23   28   16   32 | | | | | |

b 15   36   23   28   31 | | | | | |

c 23   32   34   43   29 | | | | | |

d 41   14   32   43   33 | | | | | |

e 47   37   45   38   49 | | | | | |

2 Write each set of numbers in order from greatest to smallest.

a 14   19   21   32   23 | | | | | |

b 32   50   34   45   46 | | | | | |

c 19   21   38   23   14 | | | | | |

d 12   23   24   50   48 | | | | | |

e 45   35   24   37   41 | | | | | |

➡ *Pupil Book page 19*

2D and 3D shapes

3D shapes

These are 3D shapes.

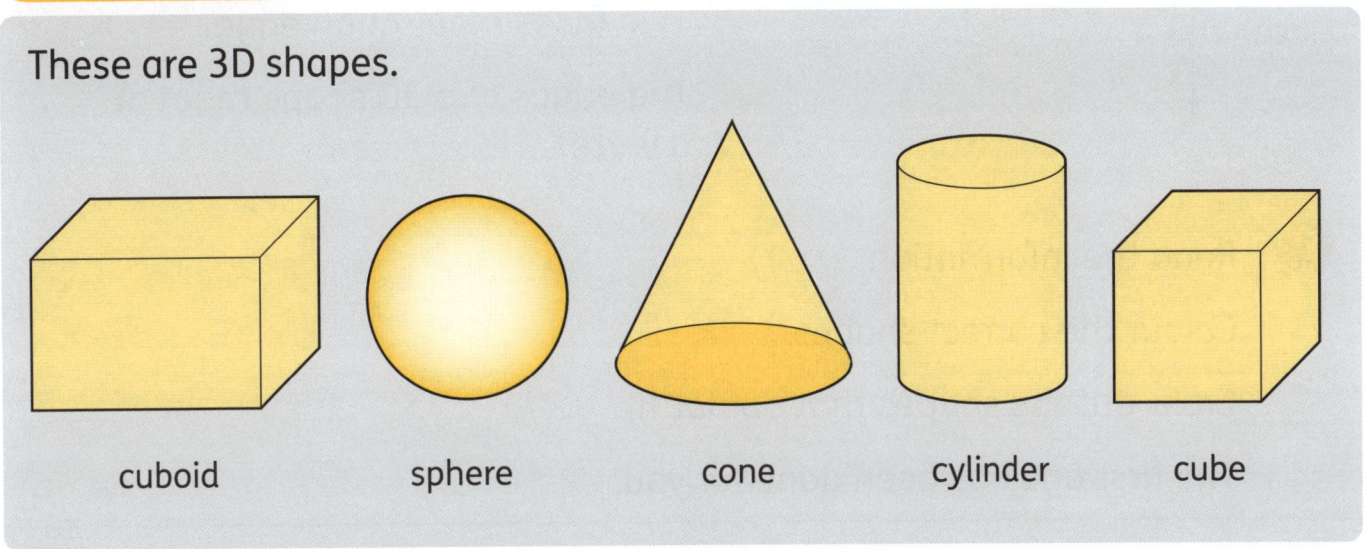

cuboid   sphere   cone   cylinder   cube

1 Draw a line to match each shape to an everyday object that is the same shape.

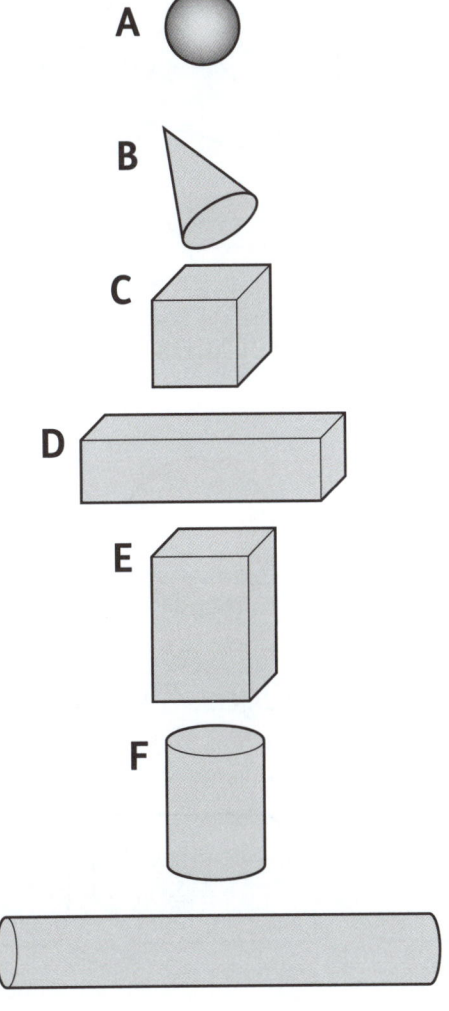

A

B

C

D

E

F

G

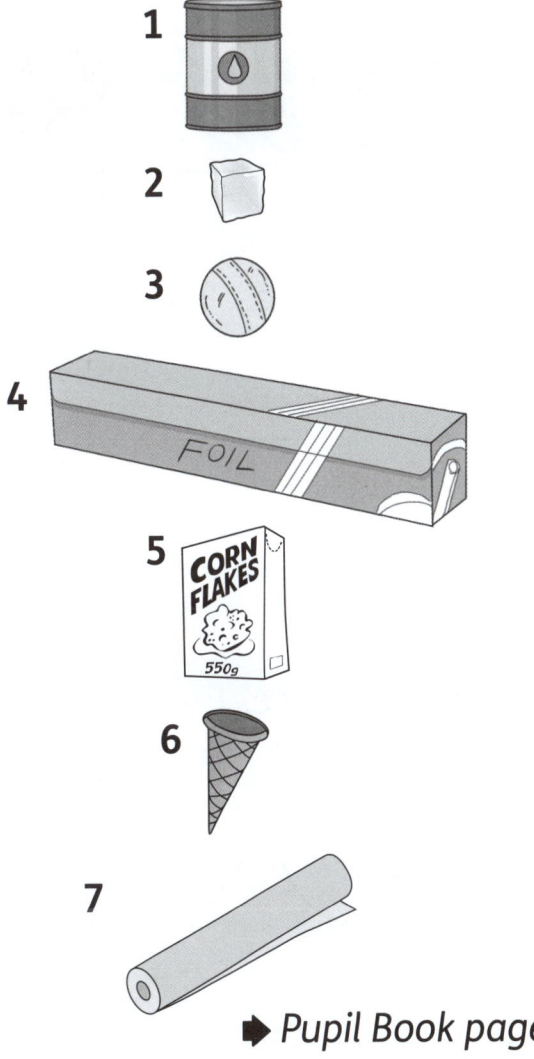

1

2

3

4 FOIL

5 CORN FLAKES 550g

6

7

➤ *Pupil Book page 22*

Which shape is it?

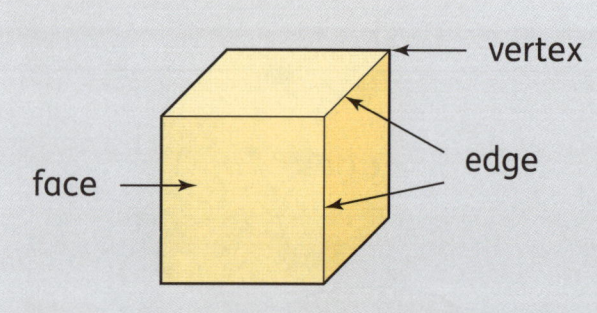

vertex

edge

face

A flat surface on a 3D shape is called a face.

Two faces meet at an edge.

The edges of a 3D shape meet at a vertex.

1 Read the information.

Colour the correct shapes.

Cross out the shapes that do not fit.

The first one has been done for you.

a 1 curved surface, 0 faces, 0 edges, 0 vertices

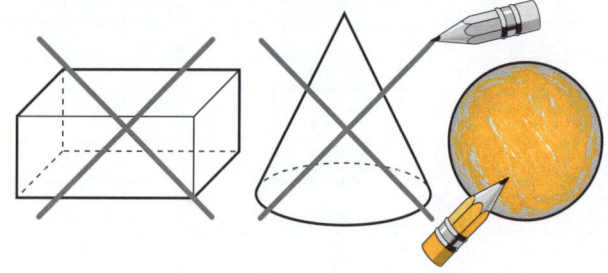

b 6 faces, 12 edges, 8 vertices

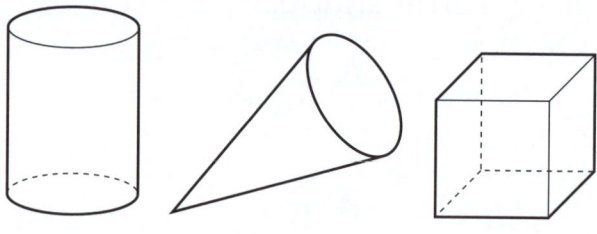

c 6 faces, 12 edges, 8 vertices

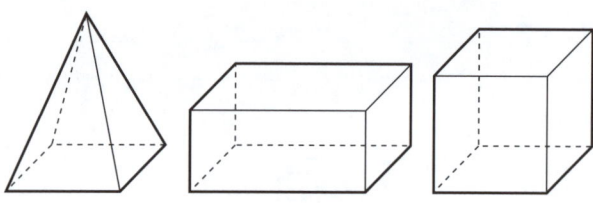

d 1 curved surface, 2 faces, 2 edges, 0 vertices

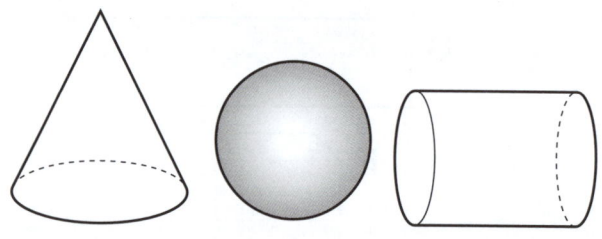

e 1 curved surface, 1 face, 1 edge, 1 vertex

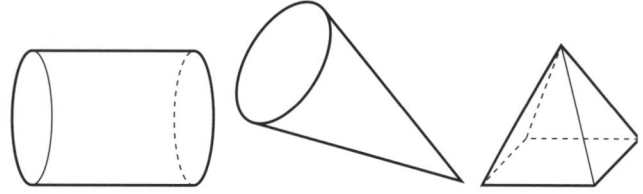

f 5 faces, 8 edges, 5 vertices

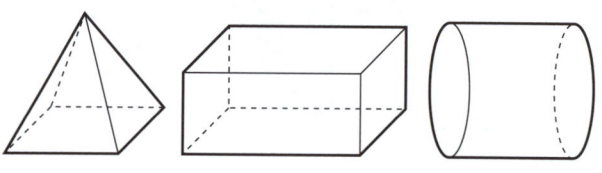

➡ *Pupil Book page 24*

Properties of 3D shapes

1 Complete this chart about the properties of 3D shapes.

3D shape	Faces	Curved surfaces	Edges	Vertices (corners)
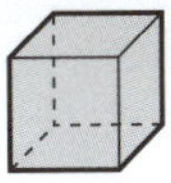 cube	6 squares	0		8
 sphere		1	0	
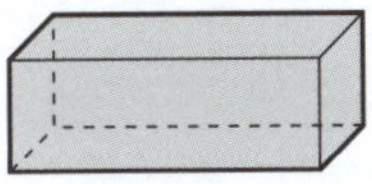 cuboid	6 rectangles			
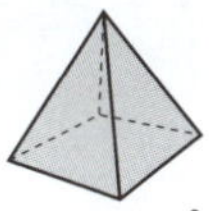 pyramid				
cylinder		1		
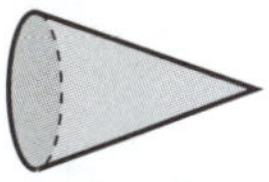 cone	1 circle			
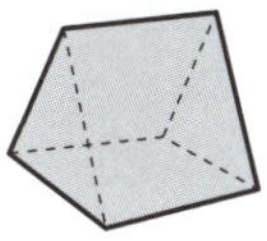 triangular prism				6

2 **a** How are a cube and cuboid the same? _____

_____

b How are they different? _____

_____

➡ *Pupil Book page 24*

Sides and vertices

1 Join dots to make the shapes.

How many sides does a square have?

4

How many vertices does a square have?

4

---

a Draw two different squares.

How many sides does a square have?

How many vertices does a square have?

---

b Draw two different triangles.

How many sides does a triangle have?

How many vertices does a triangle have?

---

c Draw two different rectangles.

How many sides does a rectangle have?

How many vertices does a rectangle have?

➡ *Pupil Book page 26*

Patterns and sequences

Shape patterns

1 Complete the pattern in each picture using similar shapes.

a

b

c

d

➡ *Pupil Book page 27*

Pattern rules

1 Give each shape a letter. When the same shape appears again, give it the same letter. Then work out the pattern rule.

A B C A B C A B

The rule is: ABC

a

A B

The rule is:

b

The rule is:

c

The rule is:

2 Make your own patterns. Use any of these shapes.

Draw a core and repeat it to make your pattern.

a Core   Pattern

b Core   Pattern

➡ *Pupil Book page 29*

Number patterns

1 Complete each number pattern.

Tell a partner how each pattern works.

a

25 30 35

b
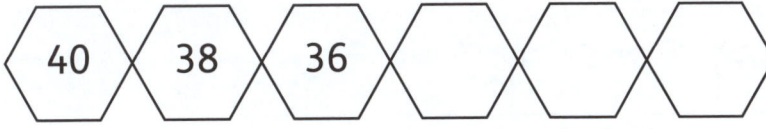
40 50 60

c
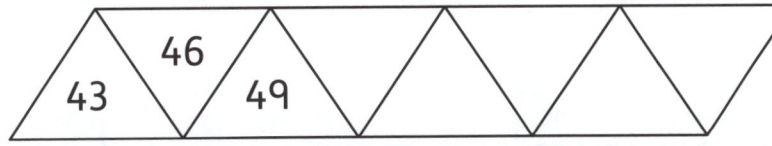
40 38 36

d
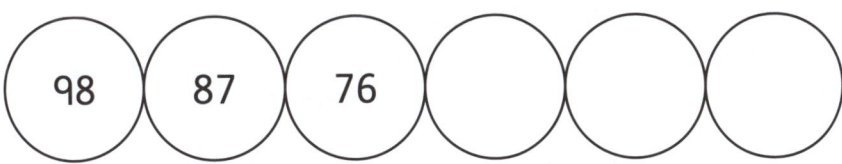
46 43 49

e

98 87 76

f

11 22 33

g
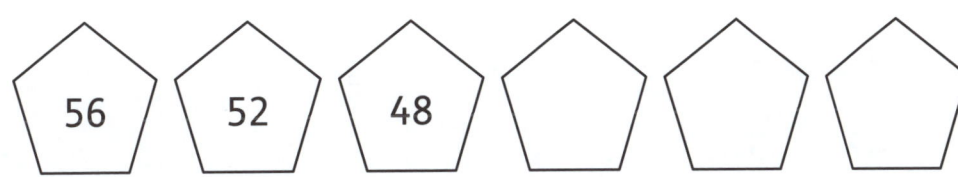
12 23 34

h
56 52 48

➡ *Pupil Book page 32*

Skip-counting patterns

1 Complete each number line.

Tell a partner what the pattern is.

a

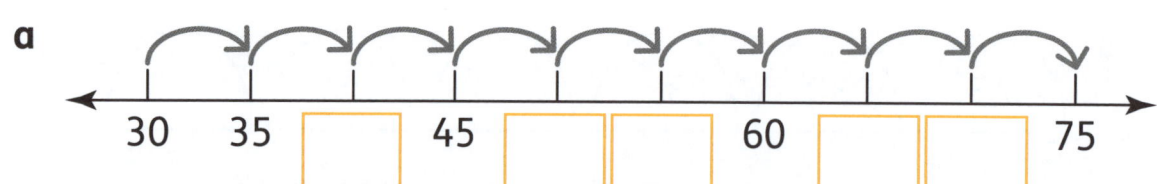

30   35   [ ]   45   [ ]   [ ]   60   [ ]   [ ]   75

b

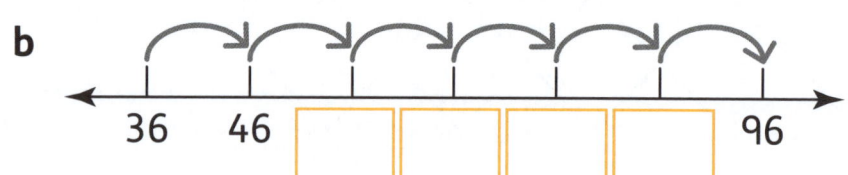

36   46   [ ]   [ ]   [ ]   [ ]   96

c

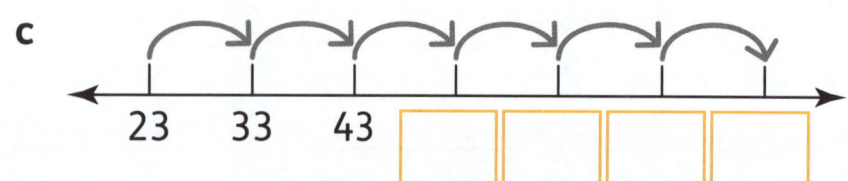

23   33   43   [ ]   [ ]   [ ]   [ ]

d

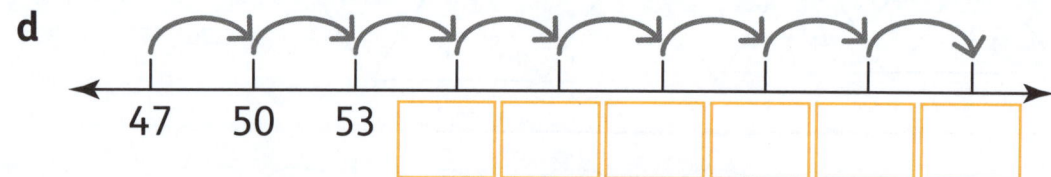

47   50   53   [ ]   [ ]   [ ]   [ ]   [ ]   [ ]

e

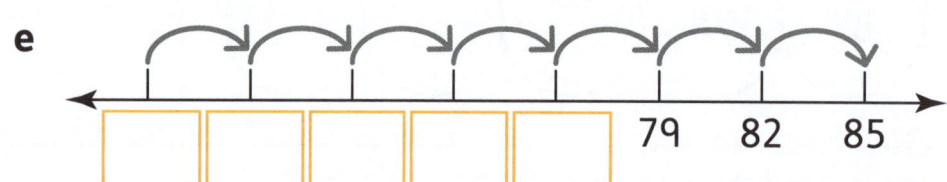

[ ]   [ ]   [ ]   [ ]   [ ]   79   82   85

---

2 Write the next three numbers in each pattern.

a 46   48   50   [ ]   [ ]   [ ]

b 85   80   75   [ ]   [ ]   [ ]

c 44   49   54   [ ]   [ ]   [ ]

d 43   47   51   [ ]   [ ]   [ ]

➡ *Pupil Book page 32*

Add and subtract

Bar models

1 Work out the missing number in each bar model.

a
20	
10	

b
5	
	2

c
20	
	2

d
7	3

e
17	3

f
10	
1	

g
1	19

h
20	
15	

2 Draw patterns to show the correct + and − sentences.

The first one has been done for you.

a
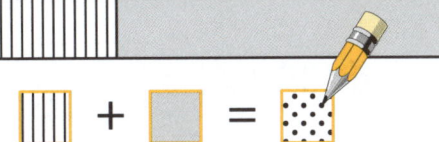

b
□ − □ = □

c
□ + □ = □

d
□ − □ = □

➡ *Pupil Book page 34*

Add tens

12 + 30

+10 +10 +10

12   22   32   42

So, 12 + 30 = 42

1	2	3	4	5	6	7	8	9	10
11	12	13	14	15	16	17	18	19	20
21	22	23	24	25	26	27	28	29	30
31	32	33	34	35	36	37	38	39	40
41	42	43	44	45	46	47	48	49	50
51	52	53	54	55	56	57	58	59	60
61	62	63	64	65	66	67	68	69	70
71	72	73	74	75	76	77	78	79	80
81	82	83	84	85	86	87	88	89	90
91	92	93	94	95	96	97	98	99	100

1 Count on in tens to find the missing numbers in these additions.

You can use the 100 chart to help you.

a 19 + 20 = ☐

b 13 + 30 = ☐

c 25 + 40 = ☐

d 39 + 40 = ☐

e 23 + 50 = ☐

f 41 + 40 = ☐

g 23 + ☐ = 53

h 18 + ☐ = 48

➡ *Pupil Book page 37*

Subtract tens

73 − 30

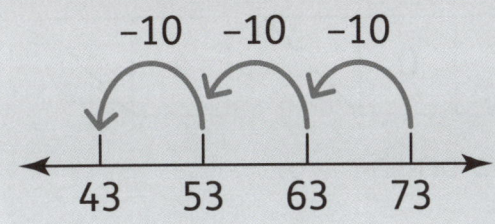

So, 73 − 30 = 43

1	2	3	4	5	6	7	8	9	10
11	12	13	14	15	16	17	18	19	20
21	22	23	24	25	26	27	28	29	30
31	32	33	34	35	36	37	38	39	40
41	42	43	44	45	46	47	48	49	50
51	52	53	54	55	56	57	58	59	60
61	62	63	64	65	66	67	68	69	70
71	72	73	74	75	76	77	78	79	80
81	82	83	84	85	86	87	88	89	90
91	92	93	94	95	96	97	98	99	100

1 Count back in tens to find the missing numbers in these subtractions.

You can use the 100 chart to help you.

a 46 − 20 =

b 67 − 40 =

c 51 − 30 =

d 67 − 50 =

e 47 − 40 =

f 39 − 20 =

g 45 − ☐ = 35

h 89 − ☐ = 69

▶ *Pupil Book page 37*

Count on in tens and ones

Remember, you can count on in tens and ones.

$32 + 24$ Think of this as 32 + | 2 | 0 | / + | 4 | /

You don't need to show the in-between numbers on the number line.

$32 + 24 =$ | 56 |

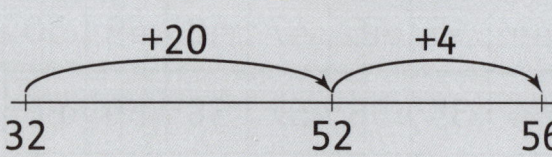

```
        +20              +4
   ⌒‾‾‾‾‾‾‾‾‾‾‾⌒    ⌒‾‾‾‾‾⌒
  32            52         56
```

1   Find the totals.

Show the jumps you do on the number lines.

a $36 + 23 =$ ☐

```
   ┼_____
   36
```

b $28 + 14 =$ ☐

```
   ┼_____
   28
```

c $38 + 14 =$ ☐

```
   ┼_____
   38
```

d $37 + 13 =$ ☐

```
   ┼_____
   37
```

e $37 + 26 =$ ☐

```
   ┼_____
   37
```

f $48 + 24 =$ ☐

```
   ┼_____
   48
```

➡ *Pupil Book page 38*

Use a 100 chart to add and subtract

You can use a 100 chart to add and subtract.

$21 + 24 = \boxed{45}$

Remember, 24 is 2 tens and 4 ones.

11	12	13	14	15	16	17	18	19	20
21	22	23	24	25	26	27	28	29	30
31	32	33	34	35	36	37	38	39	40
41	42	43	44	45	46	47	48	49	50
51	52	53	54	55	56	57	58	59	60
61	62	63	64	65	66	67	68	69	70
71	72	73	74	75	76	77	78	79	80
81	82	83	84	85	86	87	88	89	90

1 Use the 100 chart to answer the number problems.

Show the jumps for each problem in a different colour.

a $6 + 13 = \boxed{}$

b $47 - 32 = \boxed{}$

c $50 + 16 = \boxed{}$

d $23 - 12 = \boxed{}$

e $74 + 23 = \boxed{}$

f $68 - 32 = \boxed{}$

1	2	3	4	5	6	7	8	9	10
11	12	13	14	15	16	17	18	19	20
21	22	23	24	25	26	27	28	29	30
31	32	33	34	35	36	37	38	39	40
41	42	43	44	45	46	47	48	49	50
51	52	53	54	55	56	57	58	59	60
61	62	63	64	65	66	67	68	69	70
71	72	73	74	75	76	77	78	79	80
81	82	83	84	85	86	87	88	89	90
91	92	93	94	95	96	97	98	99	100

▶ *Pupil Book page 38*

Make 100

We can add tens in different ways to make 100.

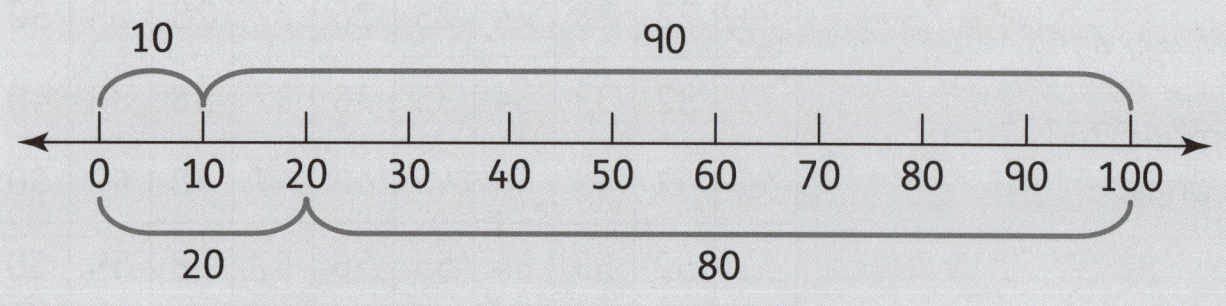

1 Use the number line to help you complete these target diagrams.

The two numbers in each section must add up to 100.

a
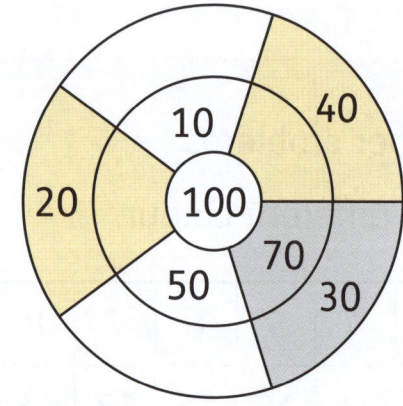

b
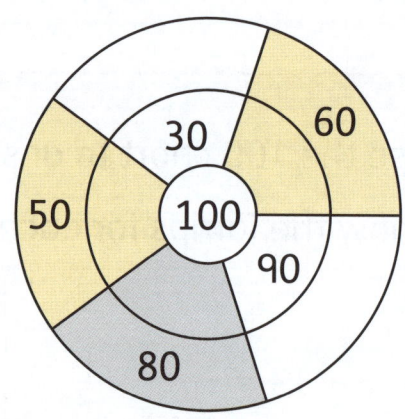

2 Complete these additions.

a 10 + 90 = ☐

b 20 + ☐ = 100

c ☐ + 70 = 100

d 40 + ☐ = 100

e ☐ + 50 = 100

f 0 + ☐ = 100

3 Now try these subtractions.

a 100 − 10 = ☐

b 100 − 90 = ☐

c 100 − 20 = ☐

d 100 − 80 = ☐

e 100 − 30 = ☐

f 100 − 70 = ☐

g 100 − 40 = ☐

h 100 − 50 = ☐

i 100 − 0 = ☐

➡ *Pupil Book page 38*

Use number facts to add

To make addition easier, you can reorder numbers and find pairs that make 10.

$4 + 9 + 6 + 1$

10 10

$= 10 + 10$
$= 20$

$3 + 9 + 6 + 1 + 4$

10 10

$= 3 + 10 + 10$
$= 23$

1   Do these additions.

Show a partner how you worked each one out.

a $3 + 7 + 9 + 1 = \boxed{}$

b $5 + 4 + 6 + 5 + 2 = \boxed{}$

c $8 + 1 + 2 + 9 = \boxed{}$

d $3 + 3 + 6 + 4 + 7 = \boxed{}$

e $3 + 6 + 5 + 4 = \boxed{}$

f $9 + 2 + 8 + 5 + 1 = \boxed{}$

g $9 + 8 + 1 + 6 + 4 = \boxed{}$

h $3 + 3 + 2 + 8 + 4 = \boxed{}$

i $8 + 7 + 3 + 7 + 2 = \boxed{}$

j $1 + 3 + 7 + 8 + 9 = \boxed{}$

➡ *Pupil Book page 39*

Add four or five numbers

You can count on to add small groups of numbers.

Use a number line, 100 chart or your fingers to help you.

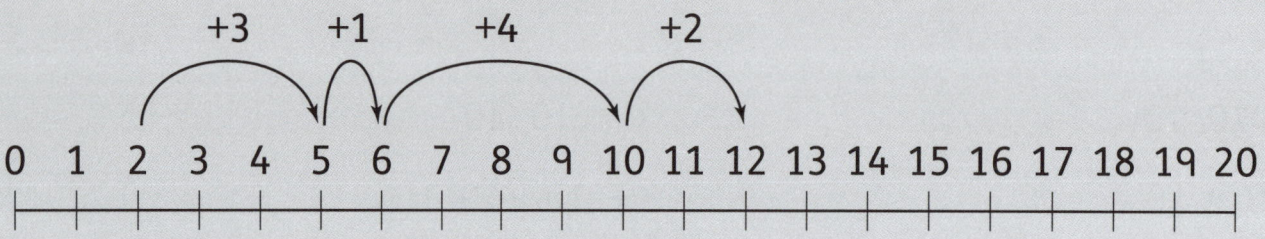

$2 + 3 + 1 + 4 + 2 = 12$

Remember, you can add in any order.

1 Add.

Show a partner how you got each answer.

a $2 + 6 + 8 + 3 =$ ☐

b $2 + 6 + 4 + 1 =$ ☐

c $1 + 7 + 3 + 9 =$ ☐

d $5 + 5 + 7 + 2 =$ ☐

e $9 + 0 + 8 + 2 =$ ☐

f $10 + 10 + 3 + 7 =$ ☐

g $6 + 3 + 2 + 8 + 4 =$ ☐

h $6 + 4 + 1 + 8 + 1 =$ ☐

i $4 + 9 + 3 + 7 + 1 =$ ☐

j $10 + 3 + 7 + 8 + 3 =$ ☐

➡ *Pupil Book page 39*

UNIT 7 Length

Measure lengths

We measure short lengths in centimetres.

Centimetres can be written as cm.

| 0 | 1 | 2 | 3 | 4 | 5 |
Centimetres (cm)

These things are all about 1 centimetre.

1 cm

A paintbrush is about 1 cm wide. 1 cm

Your finger is about 1 cm wide.

1 cm

A base-ten ones cube is 1 cm long.

1 Work with a partner.

Use base-ten ones cubes, your finger width or a ruler to measure these lengths in centimetres.

Write your own measurements in the gaps.

a My pencil is about ☐ cm long.

b A crayon is about ☐ cm long.

c My sharpener is about ☐ cm long.

d My ruler is about ☐ cm wide.

e My drinks bottle is about ☐ cm high.

f A door handle is about ☐ cm long.

➡ *Pupil Book page 43*

Estimate lengths

1 About how long are these things in real life?

Circle the best estimate.

a About 15 cm

About 150 cm

b About 10 cm

About 100 cm

c About 30 cm

About 3 cm

d About 4 cm

About 40 cm

e About 15 cm

About 30 cm

f About 30 cm

About 15 cm

g About 20 cm

About 2 cm

h About 1 cm

About 10 cm

2 Find things in your classroom that are about these lengths. Draw them.

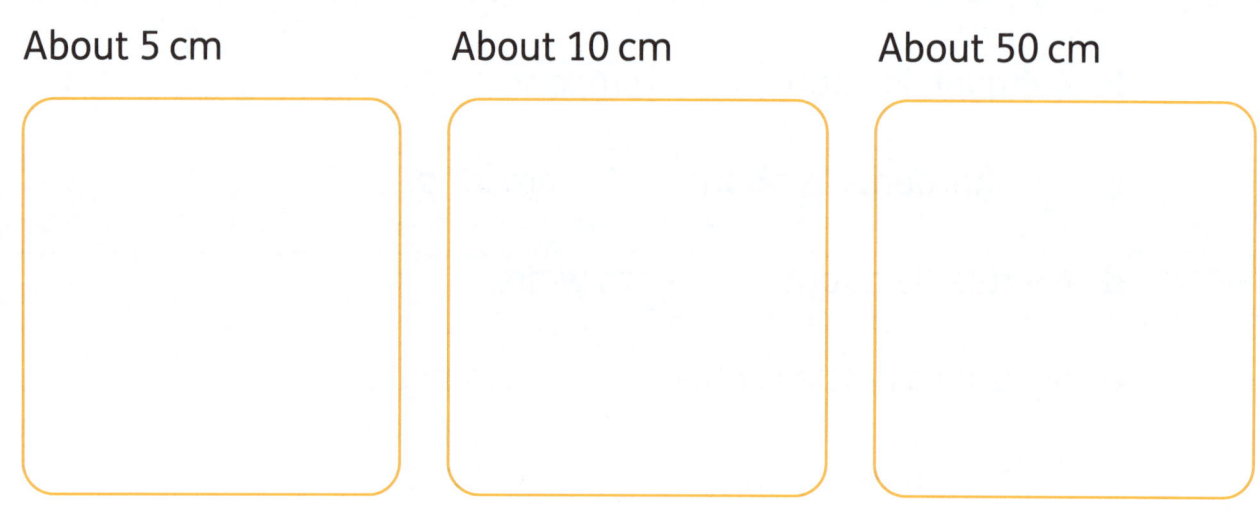

About 5 cm   About 10 cm   About 50 cm

➡ *Pupil Book page 43*

Measure in centimetres

When you use your ruler to measure in centimetres, you must measure from the 0 cm mark on the ruler.

Measure like this:

```
0   1   2   3   4
```
✔

Do not measure like this:

```
0   1   2   3   4
```
✘

1 Measure the length of each row of base-ten ones cubes. Write the answers.

a ☐ cm

b ☐ cm

c ☐ cm

d ☐ cm

e ☐ cm

2 Measure the length of each bar. Write the answers.

a ☐ cm A

b ☐ cm B

c ☐ cm C

d ☐ cm D

e ☐ cm E

f ☐ cm F

3 D is the shortest bar. Write the letters of the other bars in order from shortest to longest.

D _____ _____ _____ _____ _____ _____

➡ *Pupil Book page 44*

More centimetres

1 Use a ruler to measure the length of each straight part in centimetres.

Write the length next to each part.

Add them to find the total length.

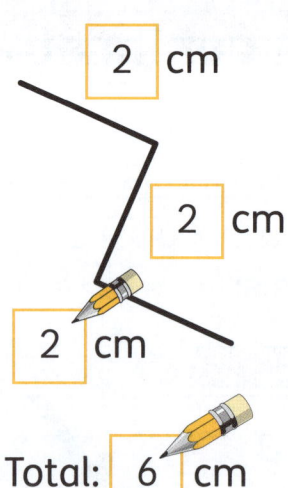

2 cm

2 cm

2 cm

Total: 6 cm

---

a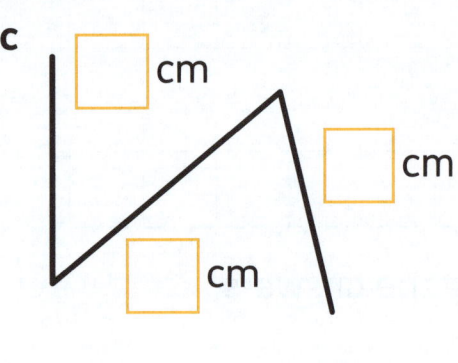

☐ cm

☐ cm

☐ cm

Total: ☐ cm

b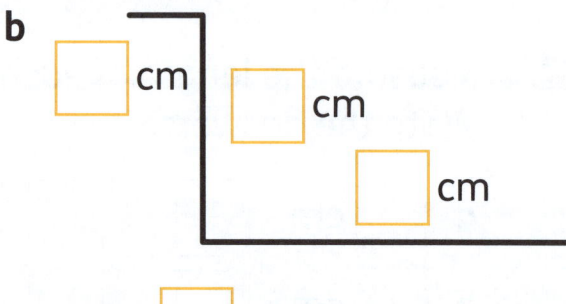

☐ cm

☐ cm

☐ cm

Total: ☐ cm

---

c

☐ cm

☐ cm

☐ cm

Total: ☐ cm

d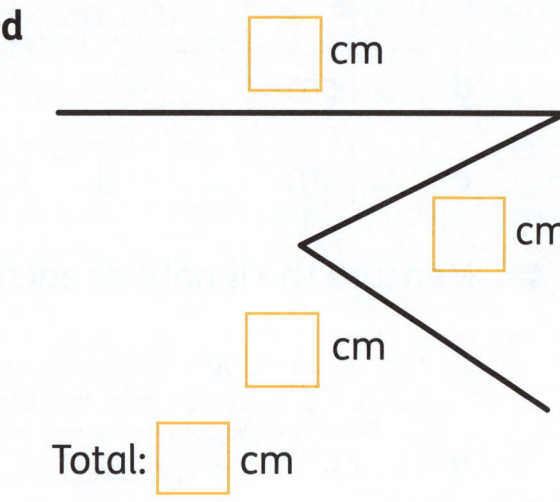

☐ cm

☐ cm

☐ cm

Total: ☐ cm

---

2 Draw a zigzag path that has a total length of 10 cm.

Compare your path with a partner's. Are they the same?

➡ *Pupil Book page 44*

Mass

Heavier or lighter

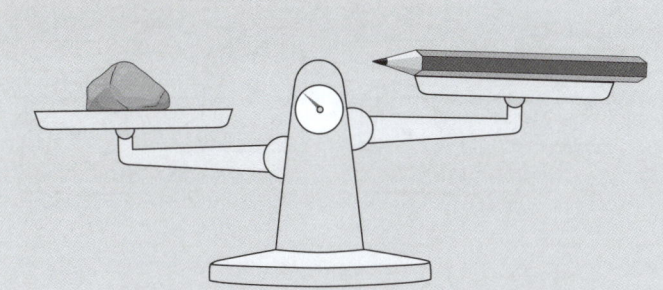

The pencil is lighter than the stone.
The stone is heavier than the pencil.

The heavier side goes down.
The lighter side goes up.

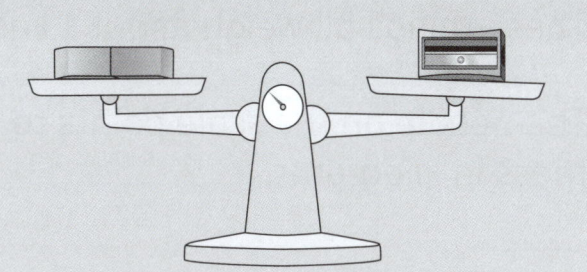

The sharpener and the eraser are equal in mass.

The pans are at the same level.

1   Draw things that are heavier or lighter for each set of scales.

a

b

c

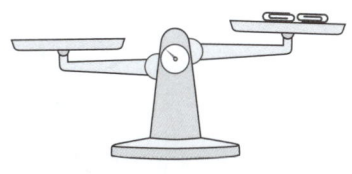

d

e

f

▶ *Pupil Book page 46*

Kilograms

We can measure mass in kilograms.

Kilograms can be written as kg.

These things all weigh about 1 kg.

The needle on the scale points to the mass in kilograms.

1 Read the mass on each scale to the nearest kilogram.

Write the mass.

a

⬜ kg

b

⬜ kg

c

⬜ kg

d

⬜ kg

e

⬜ kg

f

⬜ kg

➡ *Pupil Book page 46*

Measuring kilograms

1 Find things to weigh.

Use 1 kg weights.

Make the balance scales look like these.

Draw the thing you measured on the scales.

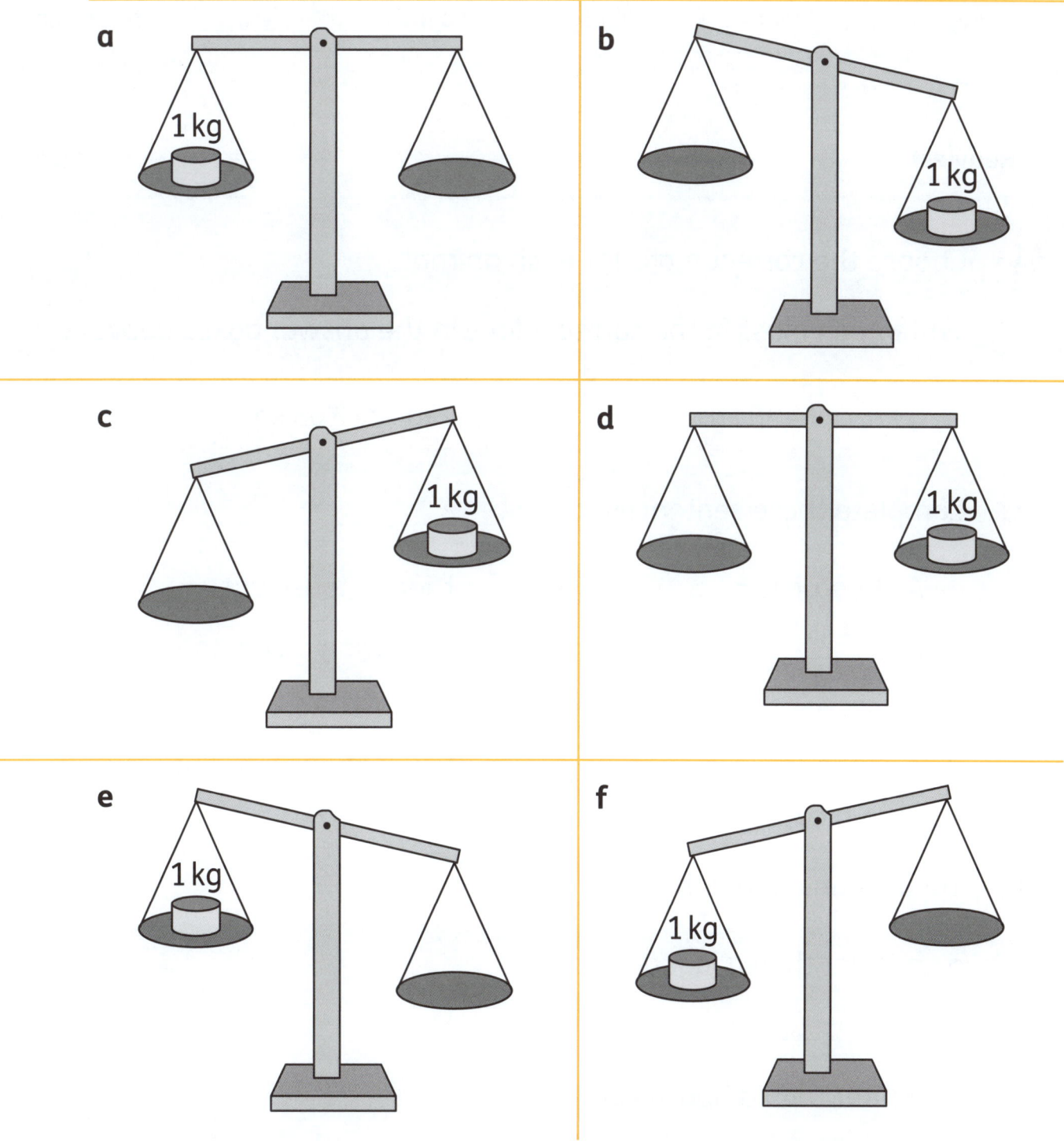

a   **b**

c   **d**

e   **f**

➡ *Pupil Book page 47*

Animal masses

These animals are shown in order from heaviest to lightest.

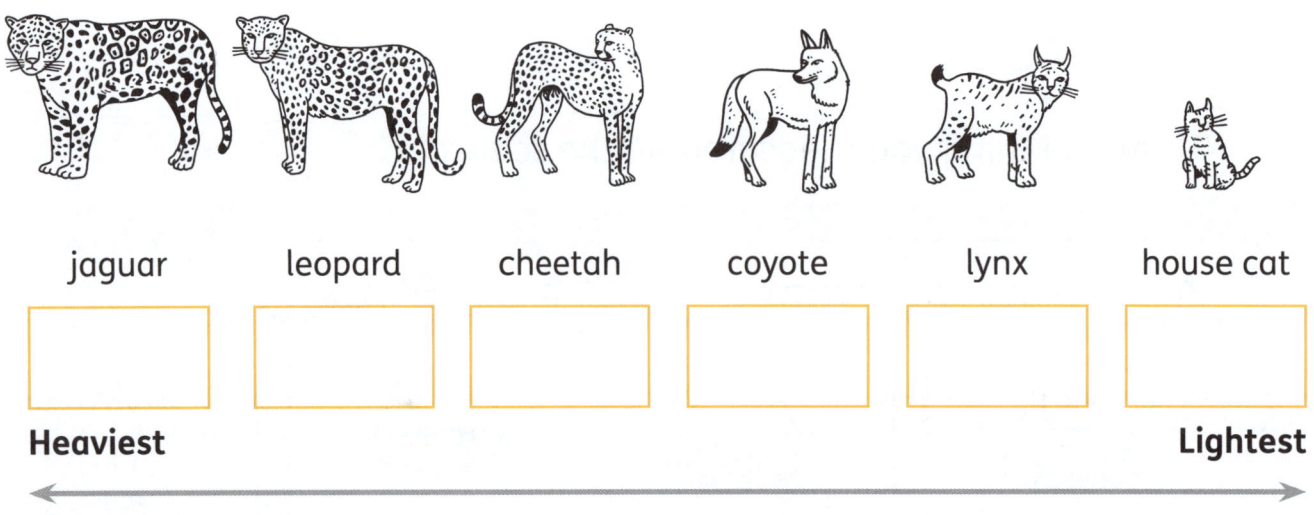

jaguar	leopard	cheetah	coyote	lynx	house cat

Heaviest   **Lightest**

1 Choose the correct mass for each animal.

Write each mass in the correct place in the answer boxes above.

| 5 kg | 90 kg | 15 kg | 20 kg | 65 kg | 45 kg |

2 Complete these sentences.

a A cheetah is _____ than a house cat.

b A _____ is 20 kg heavier than a cheetah.

c A _____ weighs less than 10 kg.

d A _____ is heavier than a house cat, but lighter than a coyote.

3 Think of an animal that is:

a heavier than a jaguar

_____

b lighter than a house cat.

_____

➡ Pupil Book page 47

Lists and tables

Sort fruits and vegetables

Apples	✓✓✓	3
Bananas	✓✓	2
	Total	5

1 Put a tick (✓) in the table for each fruit and vegetable.

Count the ticks. Write the total.

Apples		
Bananas		
Oranges		
Carrots		
Onions		
Potatoes		
	Total	

➡ *Pupil Book page 51*

Use tally tables

This is a tally table.

A tally mark is a little line.

Dogs	ЖЖ ЖЖ ЖЖ ЖЖ III	23
Cats	ЖЖ ЖЖ ЖЖ III	18

You draw a tally mark for each thing you count like this:

I = 1   II = 2   III = 3   IIII = 4

When you get to 5, you draw a line through the tally marks like this: ЖЖ

You count in groups of 5. Then you add any left-over tally marks.

ЖЖ   ЖЖ   ЖЖ   ЖЖ   III
5   10   15   20   23

You write the total in the table.

1 Draw a tally mark in the table for each shape.

Write the total for each shape.

Shape	Tally	Total
△		
○		
□		

2 Complete this tally table to show how many dogs and cats belong to pupils in your class.

Animal	Tally	Total
Dogs		
Cats		

➡ *Pupil Book page 52*

Sort numbers

1   Write the numbers in the correct places in each table.

19	5	11	12	8	7	3	18	20	17	15	1

Even	Not even

Less than 10	Not less than 10

	Even	Not even
Less than 10		
Not less than 10		

➡ *Pupil Book page 54*

Sort shapes

Here are some shapes.

We can use a Carroll diagram to sort the shapes, like this:

	Grey	Not grey
Round ◯	●	◯
Not round ⊗	▪ ▪ ▮ ▪	▭ △ ▽ ▫

1 Now sort the shapes a different way.

Draw the shapes in the correct spaces.

Colour them correctly.

	Grey	Not grey
Square ▫		
Not square ⊠		

➡ *Pupil Book page 54*

Pictograms

Janika asked her classmates about their favourite art materials.

Favourite art materials

Material	Number of pupils
chalks	☺ ☺ ☺
pencils	☺ ☺ ☺ ☺
wax crayons	☺ ☺ ☺
markers	☺ ☺ ☺ ☺ ☺ ☺
paints	☺ ☺ ☺ ☺ ☺ ☺ ☺ ☺

Key: ☺ = 1 pupil

1   Use the pictogram to help you answer the questions.

   a [ ] pupils chose pencils.

   b [ ] more pupils chose markers than pencils.

   c   Most pupils chose _____.

   d   Equal numbers of pupils chose _____ and _____.

2   Pick three ice cream flavours. Ask ten pupils to choose their favourite of those three. Use the answers to draw a pictogram.

Flavour	Number of pupils

Key: 🍦 = 1 pupil

➡ *Pupil Book page 58*

Draw a block diagram

1 Ask ten pupils in your class to tell you their favourite day of the week.

Put a tick (✓) in the table for each answer.

Favourite day of the week	Number of pupils
Monday	
Tuesday	
Wednesday	
Thursday	
Friday	
Saturday	
Sunday	

2 Use your table to complete the block diagram.

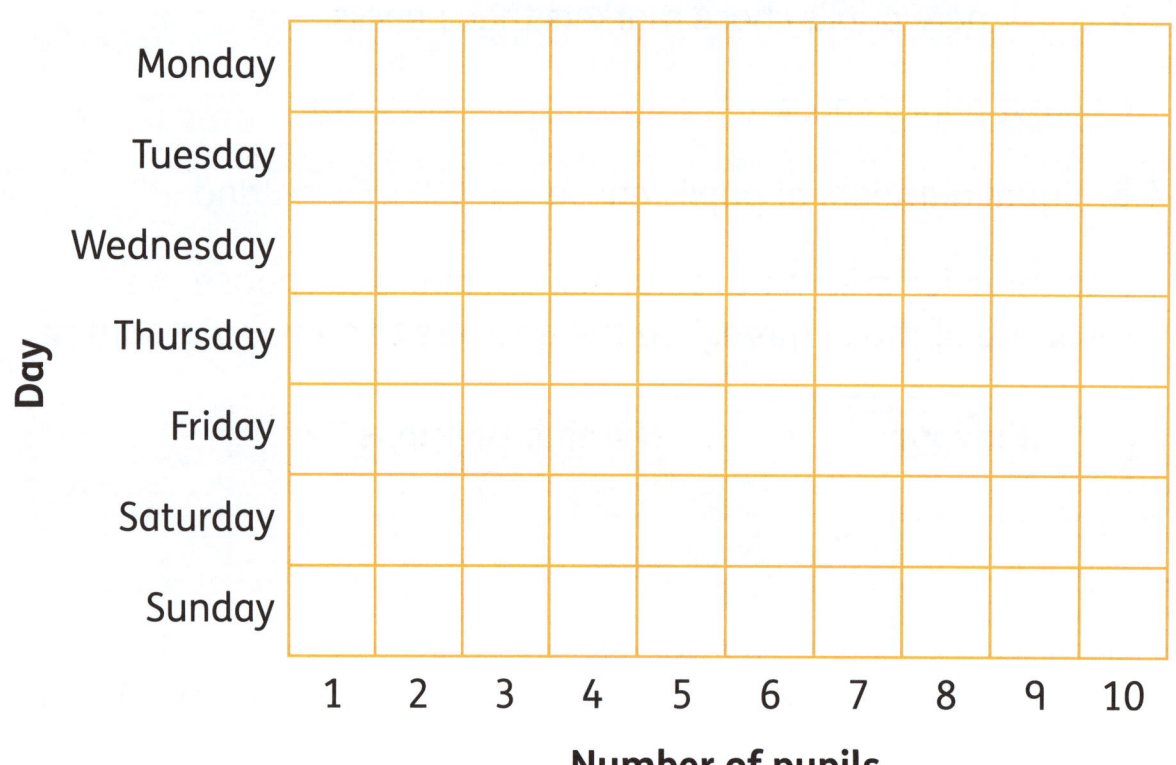

Favourite days of the week

Pupil Book page 59

Read a block diagram

Dudu asked her friends which fruit they like best.

She made this tally table to show their answers.

Our favourite fruit		
apple	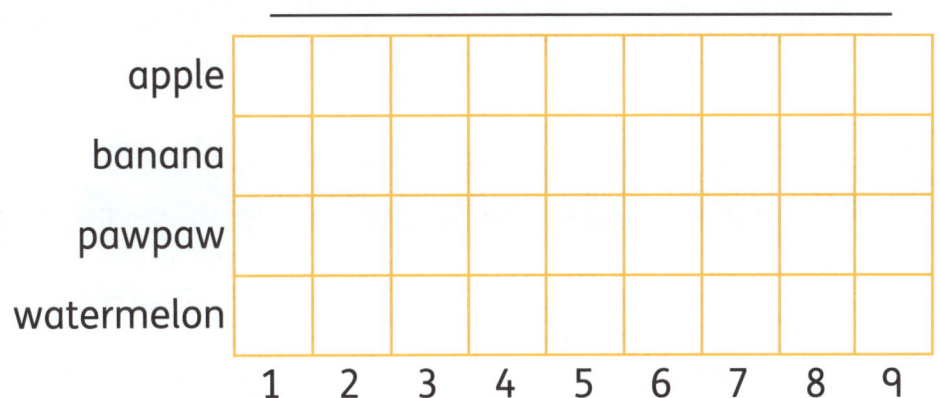	III
banana		IIII II
pawpaw		IIII I
watermelon		II

1 Complete the block diagram to show this information.

apple									
banana									
pawpaw									
watermelon									

1 2 3 4 5 6 7 8 9

2 Use your diagram to answer the questions.

   a Which fruit is the most popular? _____

   b How many pupils chose watermelon? ☐

   c How many pupils altogether chose banana and apple?

   ☐ + ☐ = ☐

   d How many more pupils chose pawpaw than chose apple?

   ☐ – ☐ = ☐

➡ *Pupil Book page 59*

Multiply

Group in different ways

1 Group the things in different ways.

Write how many groups.

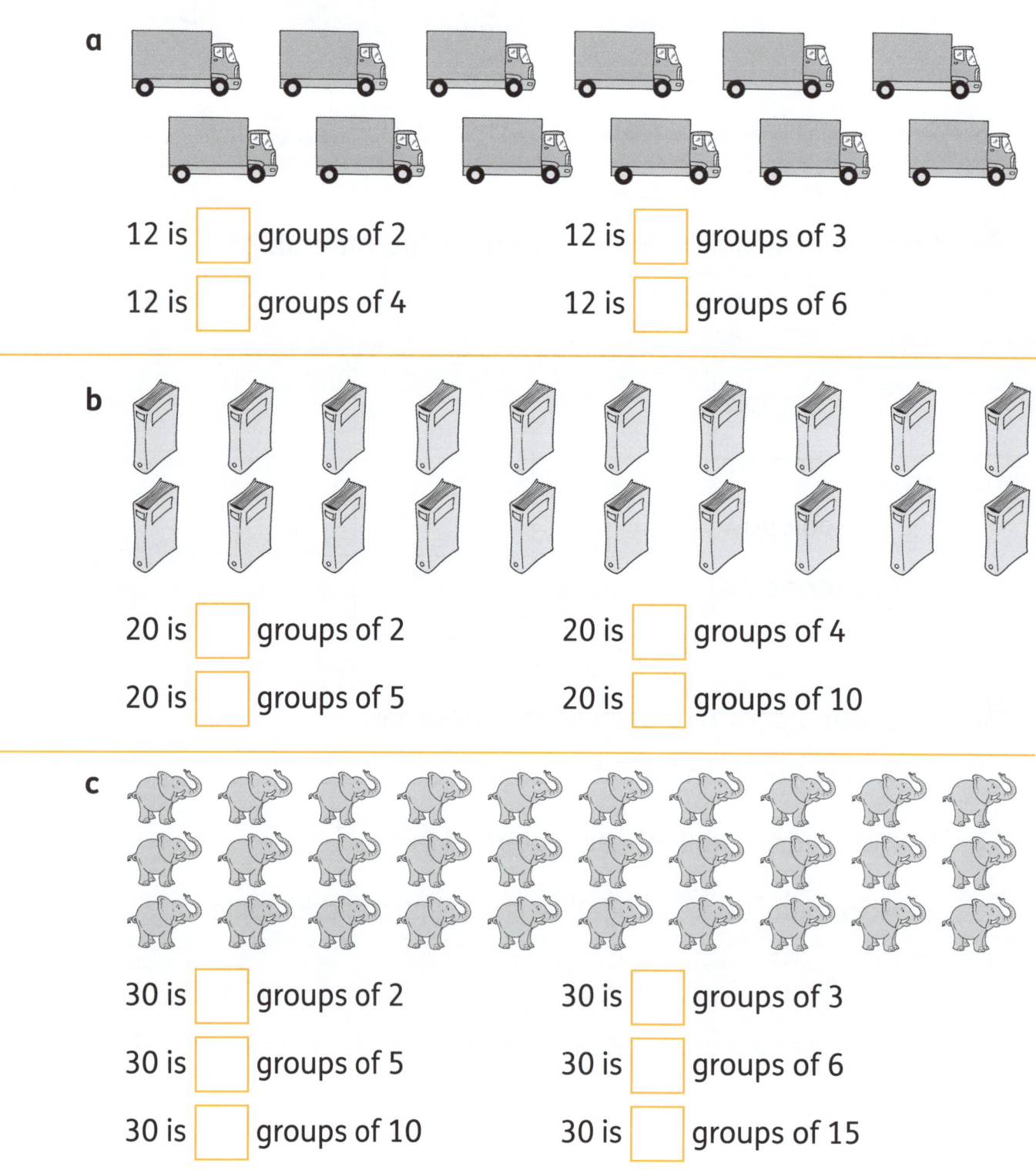

a

12 is ☐ groups of 2   12 is ☐ groups of 3

12 is ☐ groups of 4   12 is ☐ groups of 6

b

20 is ☐ groups of 2   20 is ☐ groups of 4

20 is ☐ groups of 5   20 is ☐ groups of 10

c

30 is ☐ groups of 2   30 is ☐ groups of 3

30 is ☐ groups of 5   30 is ☐ groups of 6

30 is ☐ groups of 10   30 is ☐ groups of 15

➡ *Pupil Book page 60*

Add groups

There are 3 cards of buttons.

Each card has 4 buttons.

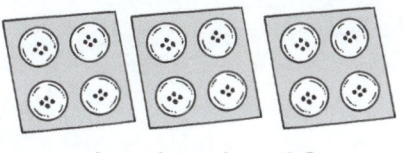

$4 + 4 + 4 = 12$

The × sign means multiply, or times.

| 3 | groups of | 4 | = | 12 |

| 3 | × | 4 | = | 12 |

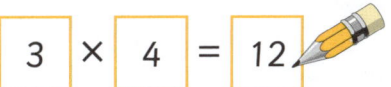

We say: '3 times 4 equals 12.'

---

1 Complete the number sentences.

a

⬚ groups of ⬚ = ⬚

⬚ × ⬚ = ⬚

b

⬚ groups of ⬚ = ⬚

⬚ × ⬚ = ⬚

c

⬚ groups of ⬚ = ⬚

⬚ × ⬚ = ⬚

d

⬚ groups of ⬚ = ⬚

⬚ × ⬚ = ⬚

e

⬚ groups of ⬚ = ⬚

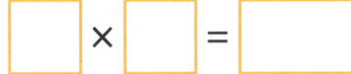

⬚ × ⬚ = ⬚

f

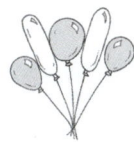  

⬚ groups of ⬚ = ⬚

⬚ × ⬚ = ⬚

➡ *Pupil Book page 61*

Rows of 2

Cindy puts 4 books on top of each other.

We can say: '1 stack of 4 is 4.'

Or: '1 times 4 is 4.'

We write: $1 \times 4 = 4$

1 Tenille arranges jelly beans in a row like this:

   a How many jelly beans are in 1 row? ☐

   b How many jelly beans are in 2 rows? ☐

   c Complete the table to show the number of jelly beans in different numbers of rows.

	1	2	3	4	5	6	7	8	9	10
Rows of 2										
Total number of beans	2	4								

2 Complete the sentences.

   a 1 row of 2 is ☐   1 times 2 = ☐ $1 \times 2 =$ ☐

   b 2 rows of 2 is ☐   2 times 2 = ☐ $2 \times 2 =$ ☐

   c 3 rows of 2 is ☐   3 times 2 = ☐ $3 \times 2 =$ ☐

   d 4 rows of 2 is ☐   4 times 2 = ☐ $4 \times 2 =$ ☐

   e 5 rows of 2 is ☐   5 times 2 = ☐ $5 \times 2 =$ ☐

➤ *Pupil Book page 61*

Arrays

1 Write two different multiplication sentences for each array.

The first one has been done for you.

a

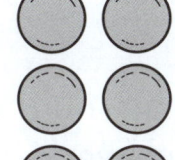

| 4 | × | 2 | = | 8 |

 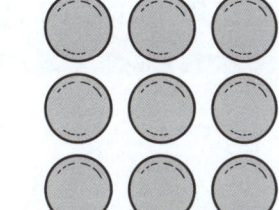

| 2 | × | 4 | = | 8 |

b

| | × | | = | |
| | × | | = | |

c

| | × | | = | |
| | × | | = | |

d

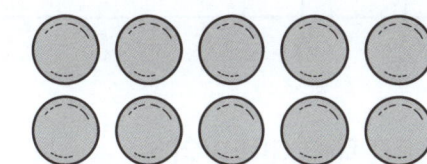

| | × | | = | |
| | × | | = | |

e

| | × | | = | |
| | × | | = | |

f

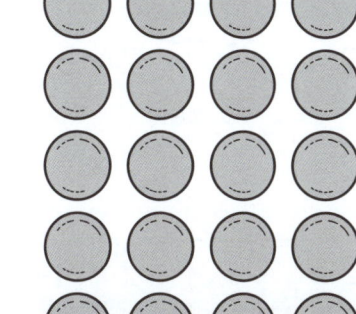

| | × | | = | |
| | × | | = | |

➡ *Pupil Book page 62*

Multiply by 2

There are 7 groups of 2 counters.

How many counters are there altogether?

You can count in groups of 2 to find the answer.

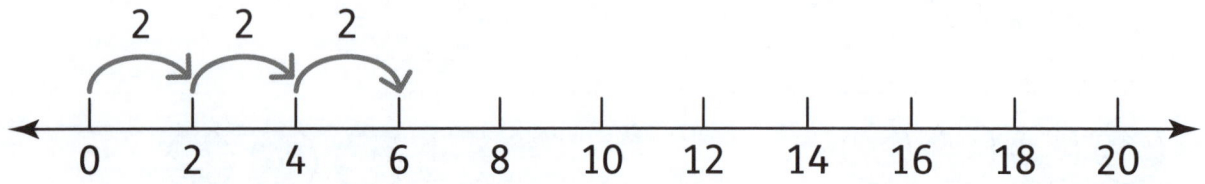

| 2 | 4 | 6 | 8 | 10 | 12 | 14 |

You can multiply to find the answer.

$$7 \times 2 = 14$$

0   2   4   6   8   10   12   14   16   18   20

1   Write the answers.

Use the number line to help you.

a $1 \times 2 =$ ☐ $2 \times 1 =$ ☐   **b** $2 \times 2 =$ ☐ $2 \times 2 =$ ☐

c $3 \times 2 =$ ☐ $2 \times 3 =$ ☐   **d** $4 \times 2 =$ ☐ $2 \times 4 =$ ☐

e $5 \times 2 =$ ☐ $2 \times 5 =$ ☐   **f** $6 \times 2 =$ ☐ $2 \times 6 =$ ☐

g $7 \times 2 =$ ☐ $2 \times 7 =$ ☐   **h** $8 \times 2 =$ ☐ $2 \times 8 =$ ☐

i $9 \times 2 =$ ☐ $2 \times 9 =$ ☐   **j** $10 \times 2 =$ ☐ $2 \times 10 =$ ☐

2   How would you work out 11×2?

Talk to a partner about your ideas.

➡ *Pupil Book page 64*

Odd and even

We can draw dots to show numbers. If we can group the dots in groups of two with none left over, the number is **even**.
If we make groups of two and one dot is left over, the number is **odd**.

1	2	3	4	5	6	7	8	9	10
odd	even	odd	even	odd	even	odd	even	odd	even

1 Write the number of dots. Circle odd or even.

a ⬜ odd even

b ⬜ odd even

c ⬜ odd even

d ⬜ odd even

e ⬜ odd even

f ⬜ odd even

2 Use counters in groups of two to make these numbers.
Write odd or even for each number.

a 12 _____ **b** 18 _____ **c** 19 _____ **d** 15 _____

e 21 _____ **f** 20 _____ **g** 17 _____ **h** 25 _____

3 Is the number 99 odd or even? _____

How do you know? _____

➡ *Pupil Book page 64*

Multiply by 5 or 10

1 Use the pictures to find the answers.

a  $3 \times 5 =$ [ ]

b 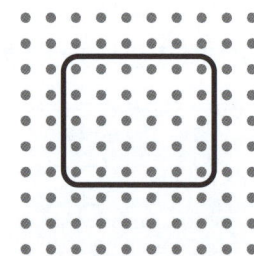 $6 \times 5 =$ [ ]

c 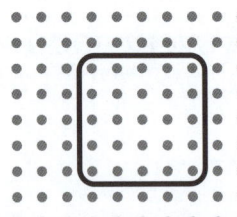 $5 \times 5 =$ [ ]

d  $7 \times 5 =$ [ ]

e  [ ] $\times 5 = 20$

f  [ ] $\times 5 = 45$

2 Shade blocks to show:

a 3×10

b 2×5

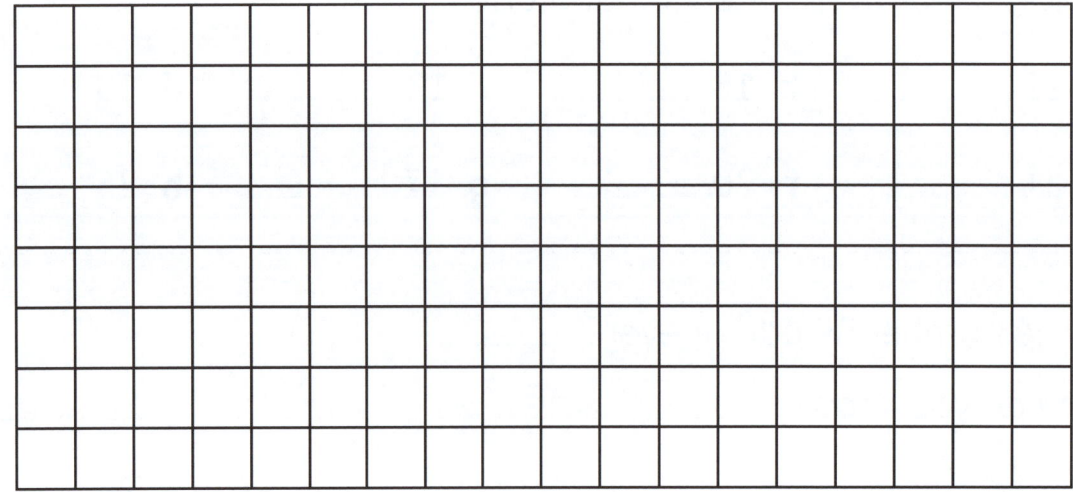

➡ *Pupil Book page 65*

Make groups

Make groups of 4.

[3] groups of [4] = [12]

You can also show this on a number line.

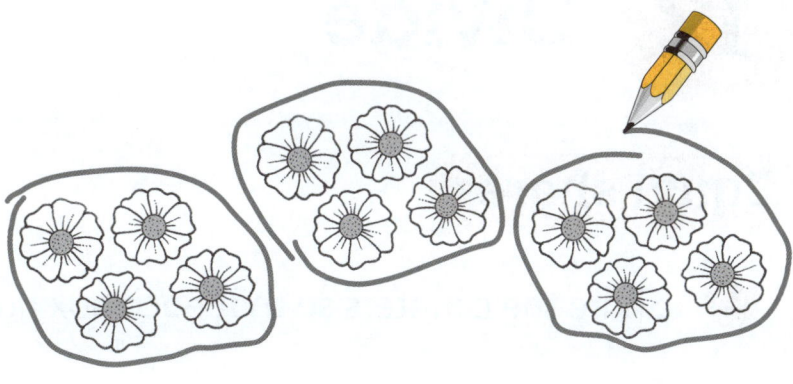

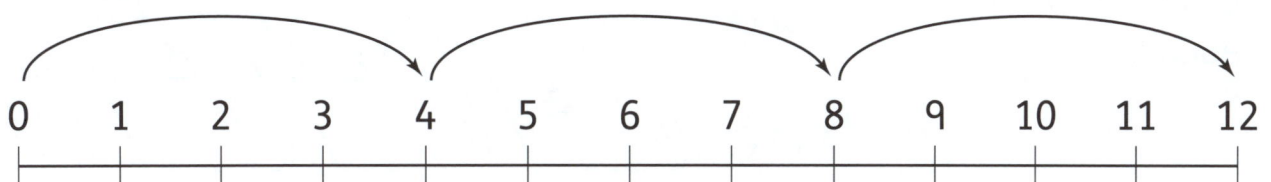

0   1   2   3   4   5   6   7   8   9   10   11   12

---

1   Circle the groups.

Complete the number sentences.

a Make groups of 2.

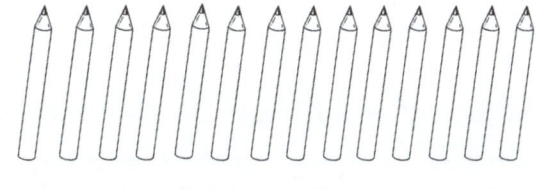

[   ] groups of 2 = [   ]

b Make groups of 3.

[   ] groups of 3 = [   ]

c Make groups of 4.

[   ] groups of 4 = [   ]

d Make 3 equal groups.

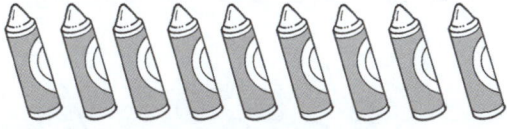

[   ] groups of [   ] = [   ]

e Make 5 equal groups.

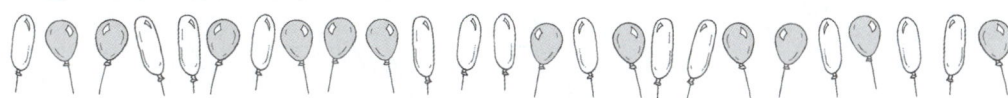

[   ] groups of [   ] = [   ]

➤ *Pupil Book page 67*

Equal shares

1 Share the counters so that each box has the same amount.

Draw the counters in each box.

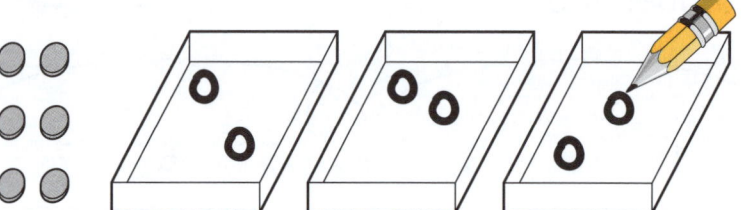

a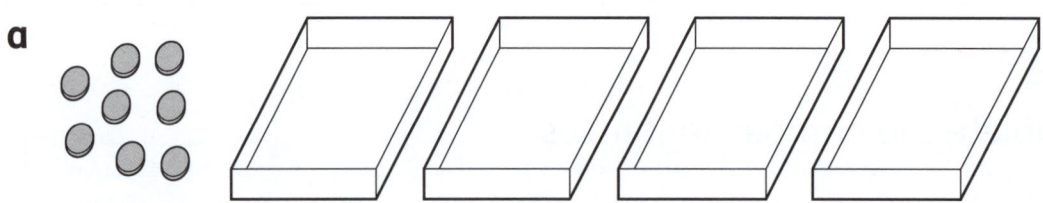

b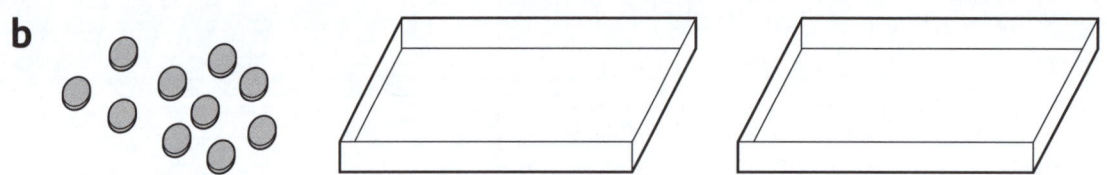

c

d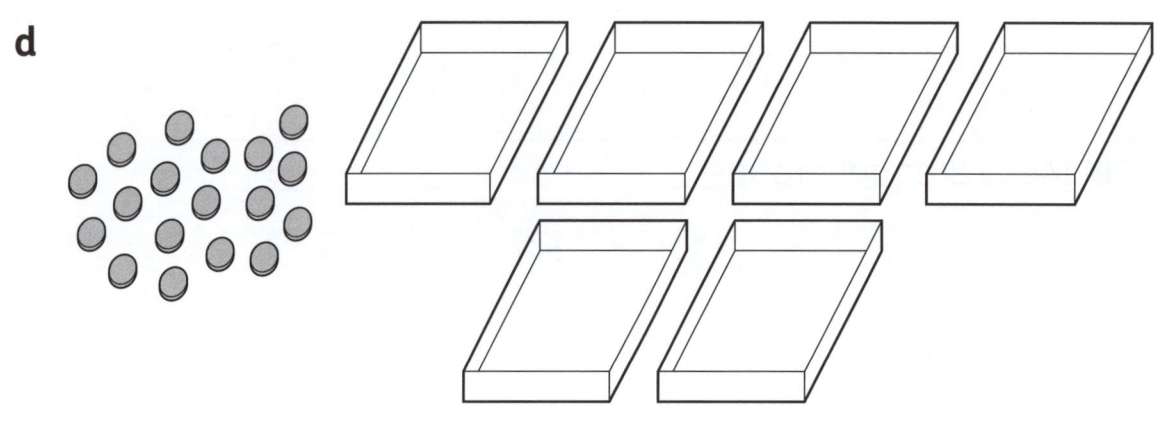

➡ *Pupil Book page 68*

Divide equally

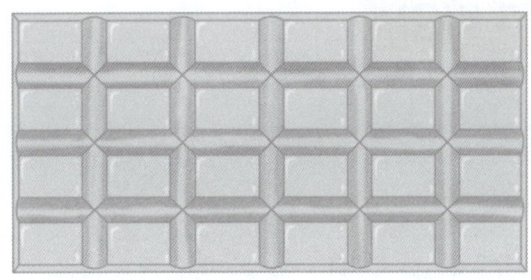

This bar of chocolate has 24 small blocks.

---

1 Draw how you would divide the chocolate to get equal shares.

Use different colours to show the shares.

a 2 equal shares

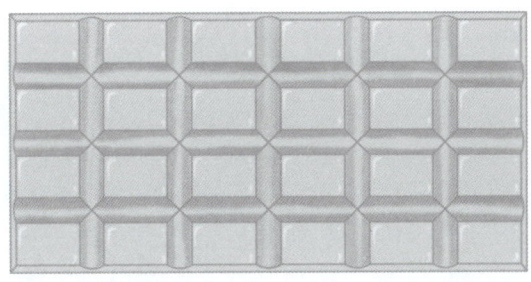

b 3 equal shares

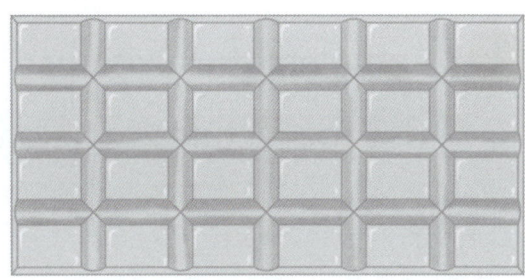

c 4 equal shares

d 6 equal shares

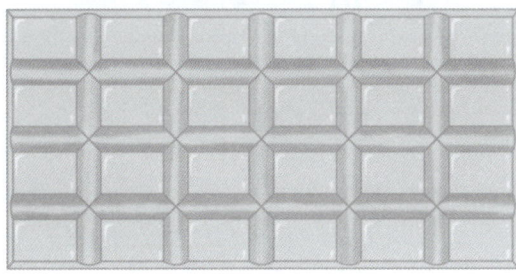

e 8 equal shares

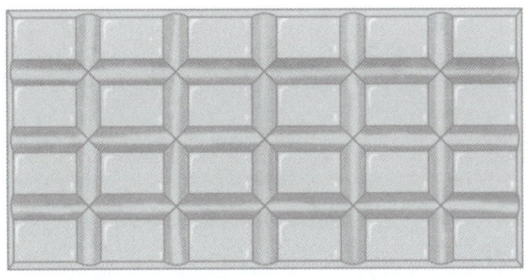

f 12 equal shares

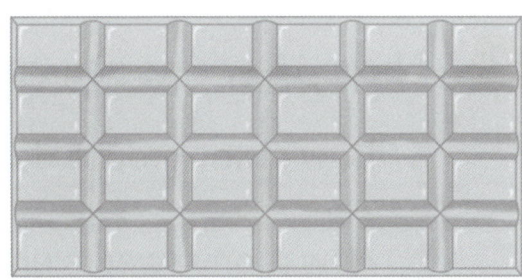

➡ *Pupil Book page 69*

The division sign

Here are 12 sweets.

How many groups of 3 can you make?

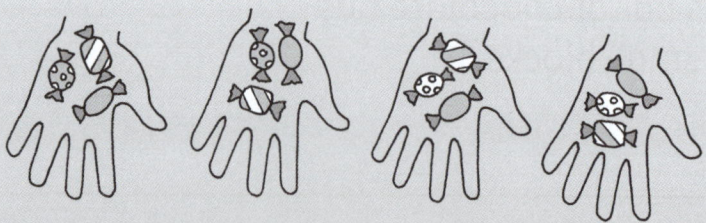

We write: 12 ÷ 3 = 4

We say: '12 divided by 3 equals 4.'

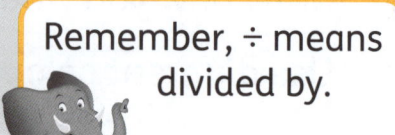

Remember, ÷ means divided by.

1 Complete the divisions.

Use the pictures to help you.

a 12 ÷ 2 = ☐   12 ÷ 6 = ☐

12 ÷ 3 = ☐   12 ÷ 4 = ☐

b 20 ÷ 2 = ☐   20 ÷ 10 = ☐

20 ÷ 4 = ☐   20 ÷ 5 = ☐

c 24 ÷ 2 = ☐   24 ÷ 4 = ☐

24 ÷ 6 = ☐   24 ÷ 8 = ☐

2 I have 15 sweets and I make groups of 2.

a How many groups do I make? ☐

b How many sweets are left over? ☐

▶ *Pupil Book page 69*

Multiplication and division facts

Division facts	Multiplication facts
$12 \div 3 = 4$	$3 \times 4 = 12$
$12 \div 4 = 3$	$4 \times 3 = 12$

If you know one fact, you can write the other three facts.

1 Complete the fact families.

a I know that: $3 \times 6 = 18$

So, I also know:

$\boxed{} \times \boxed{} = 18$

$18 \div \boxed{} = \boxed{}$

$18 \div \boxed{} = \boxed{}$

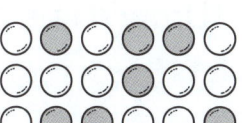

b I know that: $4 \times 5 = 20$

So, I also know:

$\boxed{} \times \boxed{} = 20$

$20 \div \boxed{} = \boxed{}$

$20 \div \boxed{} = \boxed{}$

c I know that: $3 \times 8 = 24$

So, I also know:

$\boxed{} \times \boxed{} = 24$

$24 \div \boxed{} = \boxed{}$

$24 \div \boxed{} = \boxed{}$

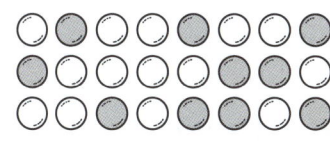

d I know that: $4 \times 10 = 40$

So, I also know:

$\boxed{} \times \boxed{} = 40$

$40 \div \boxed{} = \boxed{}$

$40 \div \boxed{} = \boxed{}$

➡ *Pupil Book page 72*

Fractions

Make equal parts

1 Colour half of each shape.

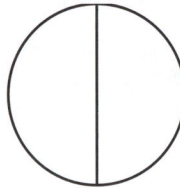  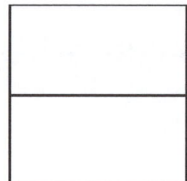 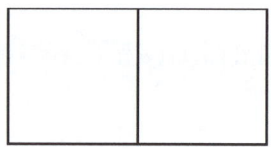

2 Draw a line to divide each shape into halves.

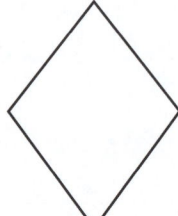    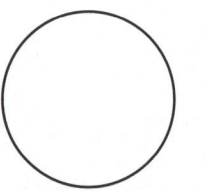 

3 Colour quarter of each shape.

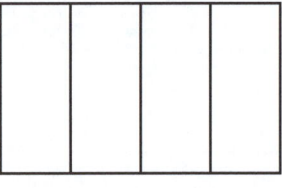 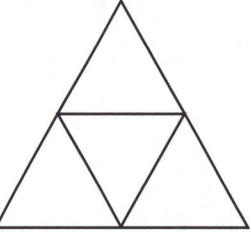 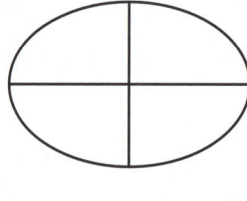 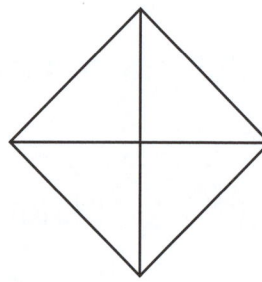

4 Draw lines to divide each shape into quarters.

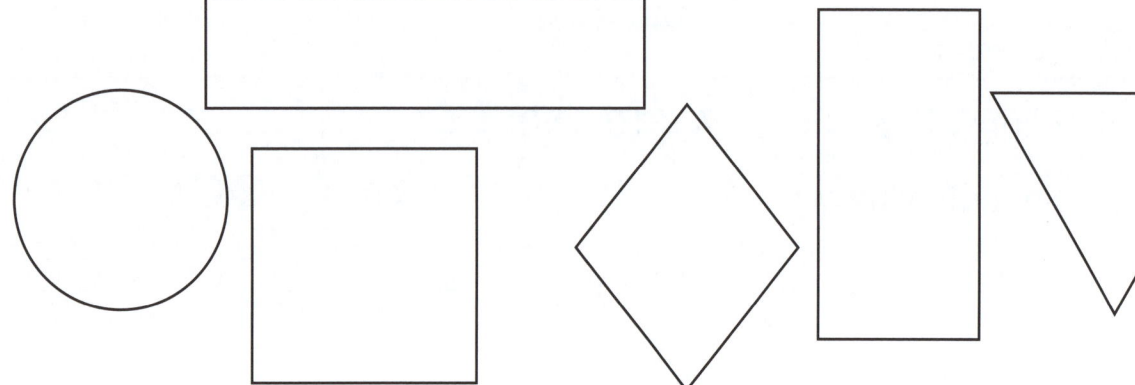

➤ *Pupil Book page 74*

Write fractions

How much of the shape is shaded?

1 out of 3 equal parts or $\frac{1}{3}$.

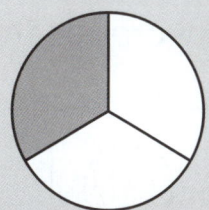

$\dfrac{1}{3}$ ← shaded part of the shape

— line meaning 'out of'

← how many equal parts make the whole

When you share an object or group into equal parts, each part is called a fraction. $\frac{1}{3}$ is a fraction.

1 How much has been shaded?

Write the fraction.

One has been done for you as an example.

a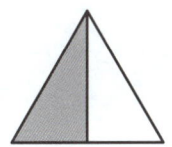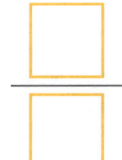

b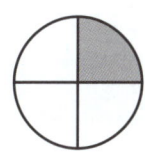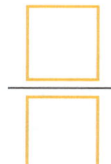

c 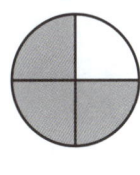 $\dfrac{3}{4}$

d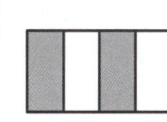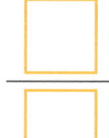

e

f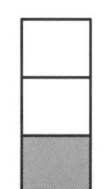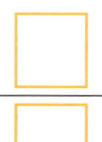

g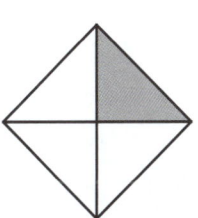

h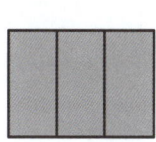

➡ *Pupil Book page 74*

Half of a group

You can make two equal groups from one bigger group.

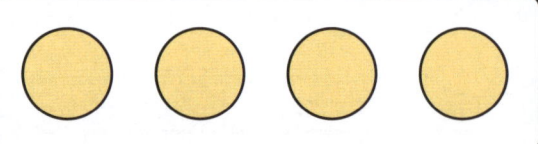

One group of 4.

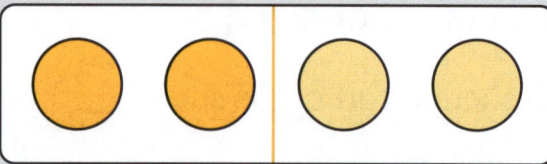

Two groups of 2.
Half of 4 is 2.

The number in each equal group is half, or $\frac{1}{2}$, the number in the bigger group.

1 Colour half the animals in each group.

Write the answers.

a $\frac{1}{2}$ of 6 is ☐

b $\frac{1}{2}$ of 10 is ☐

c $\frac{1}{2}$ of 8 is ☐

d $\frac{1}{2}$ of 4 is ☐

e $\frac{1}{2}$ of 12 is ☐

f $\frac{1}{2}$ of 2 is ☐

g $\frac{1}{2}$ of 16 is ☐

➡ *Pupil Book page 75*

Quarter of a group

You can make four equal groups from one bigger group.

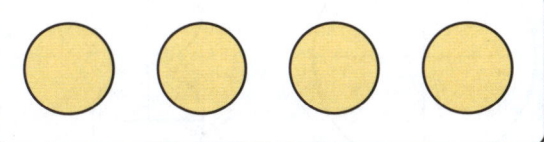

One group of 4.

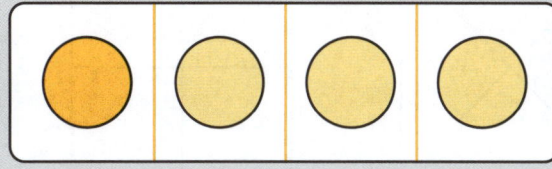

Four groups of 1.

$\frac{1}{4}$ of 4 is 1.

The number in the equal group is one-quarter, or $\frac{1}{4}$, of the number in the bigger group.

1 Colour to show one-quarter. Write the answers.

a $\frac{1}{4}$ of ☐ is ☐.

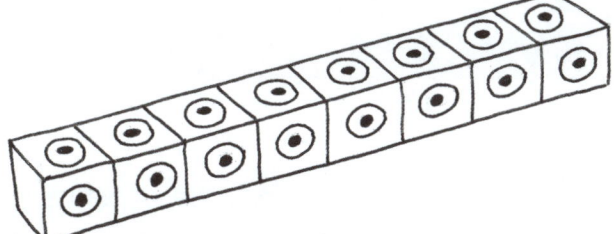

b $\frac{1}{4}$ of ☐ is ☐.

c $\frac{1}{4}$ of ☐ is ☐.

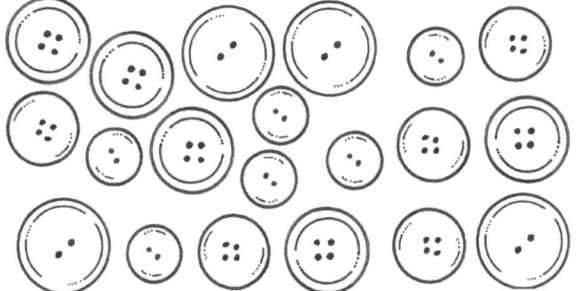

d $\frac{1}{4}$ of ☐ is ☐.

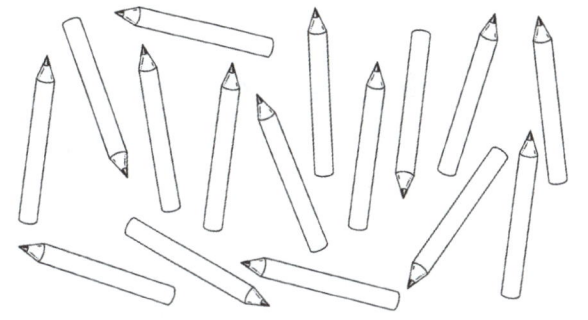

▶ *Pupil Book page 75*

Halves and quarters

1 Colour half of each shape.

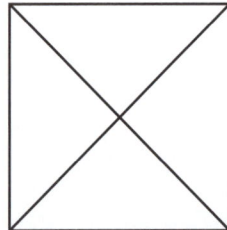 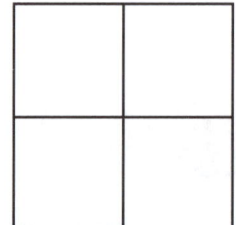 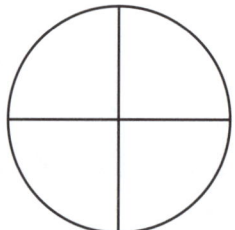 

2 Colour half of each group.

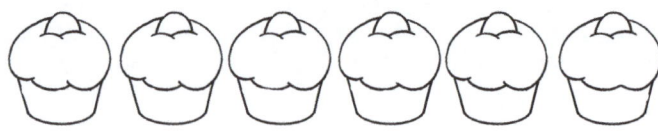 

 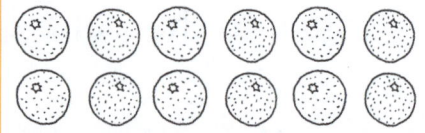

3 Colour one-quarter of each shape.

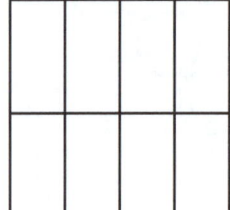 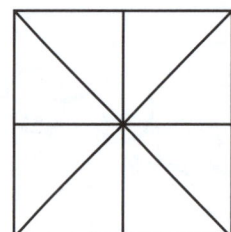 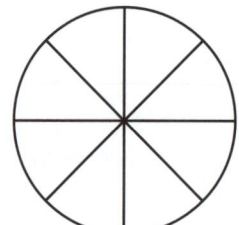 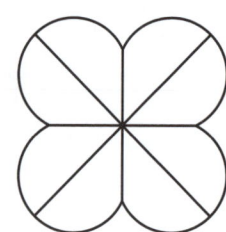

4 Colour one-quarter of each group.

  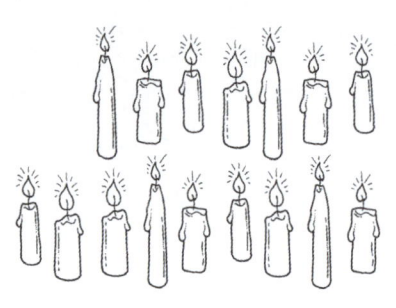

➡ *Pupil Book page 75*

Fraction problems

Salma has two cakes.

A

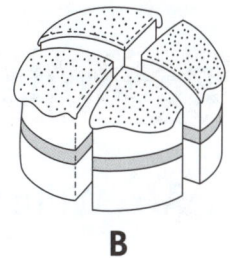

B

1 Write answers to the questions.

a How many pieces did she cut cake A into?

☐

b What fraction is each piece of cake A?

_____

c How many pieces did she cut cake B into?

☐

d What fraction is each piece of cake B?

_____

2 Salma gives some of the cake to her neighbours.

a She gives Mrs Moosa one piece of cake A.

How much is left of cake A? _____

b She gives Mr Bloom two pieces of cake B.

How much is left of cake B? _____

3 Colour these fraction bars to show what is left of each cake.

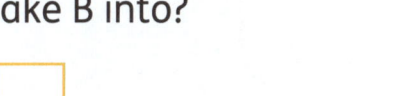

Cake A

half	
quarter	quarter

Cake B

half	
quarter	quarter

4 Complete this sentence:

Half of cake A is the same size as _____ of cake B.

➡ *Pupil Book page 78*

Seconds, minutes and hours

1 Write these times in the correct places in the table.

a 45 seconds   **b** 1 second   **c** 120 seconds

d 2 minutes   **e** 59 seconds   **f** 60 seconds

g 1 hour   **h** 1 day   **i** $\frac{1}{2}$ a second

Less than 1 minute	Exactly 1 minute	More than 1 minute

2 Write these times in the correct places in the table.

a $\frac{1}{2}$ an hour   **b** 120 minutes   **c** 1 day

d 60 minutes   **e** 60 seconds   **f** 45 minutes

g 90 minutes   **h** 62 minutes   **i** $\frac{1}{2}$ a day

Less than 1 hour	Exactly 1 hour	More than 1 hour

➡ *Pupil Book page 81*

Months of the year

1 Write these months in the correct order.

Two have been done for you.

| March | July | August | May | November |

| June | February | September | April | October |

1st _____January_____   7th _____

2nd _____   8th _____

3rd _____   9th _____

4th _____   10th _____

5th _____   11th _____

6th _____   12th _____December_____

2 Here are the dates of five pupils' birthdays.

Ling 19th January   Saleem 23rd October   John 14th May   Lean 7th November   Dwayne 18th July

a Whose birthday is first in the year? _____

b Whose birthday is in the fifth month? _____

c Which pupil has the last birthday in the year? _____

d In which month were you born? _____

➡ Pupil Book page 83

Compare times

1 Colour the equal times the same colour.

7 days

1 minute

60 seconds

1 year

2 weeks

60 minutes

1 hour

12 months

1 week

14 days

30 days

$\frac{1}{2}$ a year

1 month

6 months

2 Write each set of times in order from shortest to longest.

a 1 year   6 months   9 months   4 weeks

_____

b 40 days   20 days   1 month $\frac{1}{2}$ a month

_____

c 3 hours   40 seconds   2 minutes   1 day

_____

➡ Pupil Book page 85

Possible outcomes

Possible outcomes

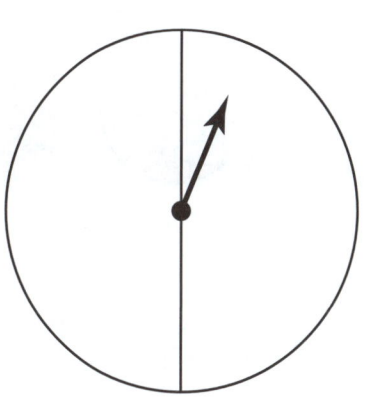

1 Colour each half of the spinner a different colour.

Then complete the sentences.

There are ☐ possible outcomes.

The spinner could land on _____ or it could land on _____ .

It is impossible for the spinner to land on _____ .

2 If something is possible, that means _____

_____ .

If something is impossible, that means _____

_____ .

3 **a** Draw something that is possible for you to do today.

b Draw something that is impossible for you to do today.

Pupil Book page 86

Regular and random patterns

1 Continue the regular patterns.

a

| 🌙 | ☀ | 🌙 | ☀ | | | |

b

| ✳ | ☁ | ☁ | | | | |

c

| 👁 | 👁 | ♡ | | | | |

d

| 1 | 2 | 3 | 1 | 2 | | |

2 Continue these to make random patterns.

a

| 1 | 2 | | | | | |

b

| ☀ | ☁ | | | | | |

c

| 👁 | 👁 | | | | | |

➡ *Pupil Book page 88*

UNIT 16

Symmetry

Line symmetry

1 Draw a line of symmetry on each shape. Use a ruler.

Talk about your answers with a partner.

a

b

c

d

e

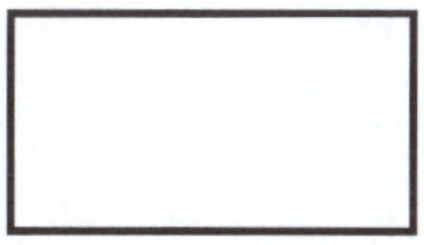

f

g

h

2 Draw a line of symmetry on each object.

One has been done for you.

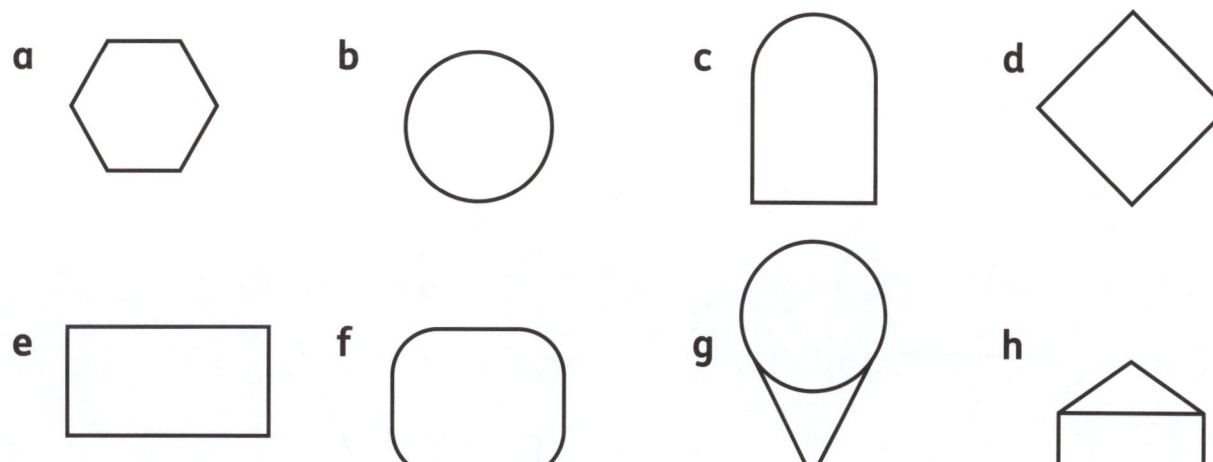

a

b

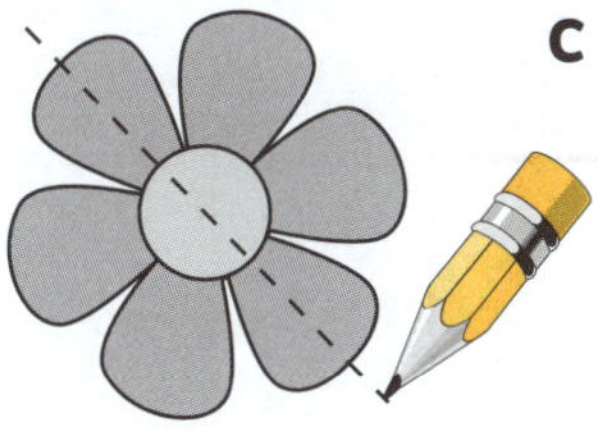

c

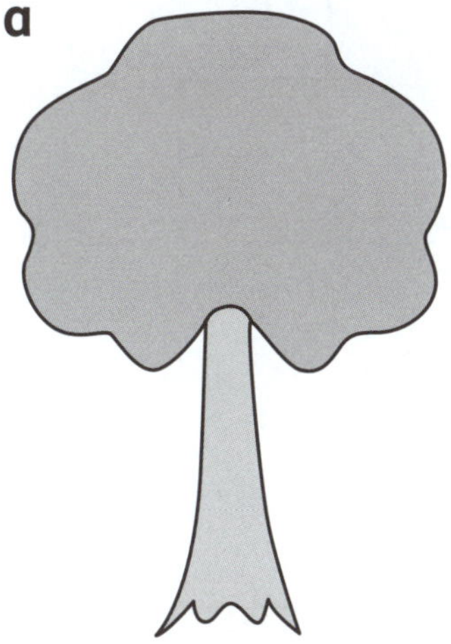

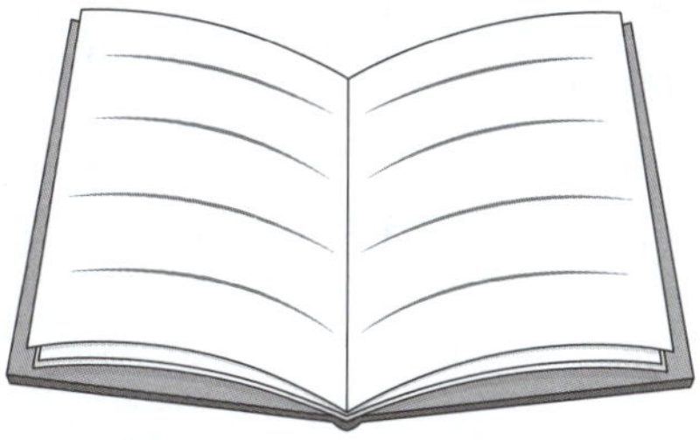

d

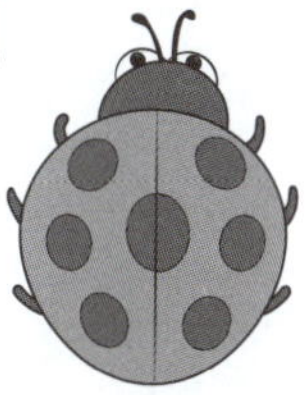

e

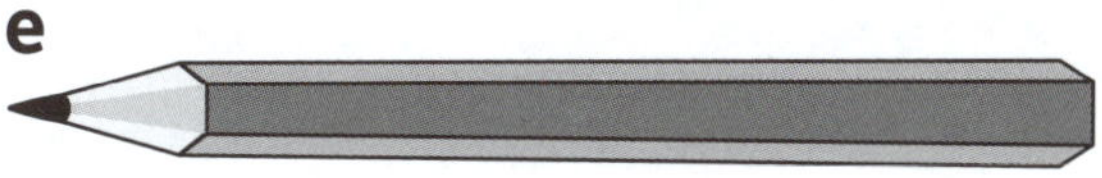

f

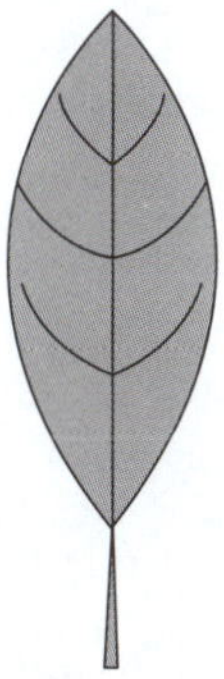

g

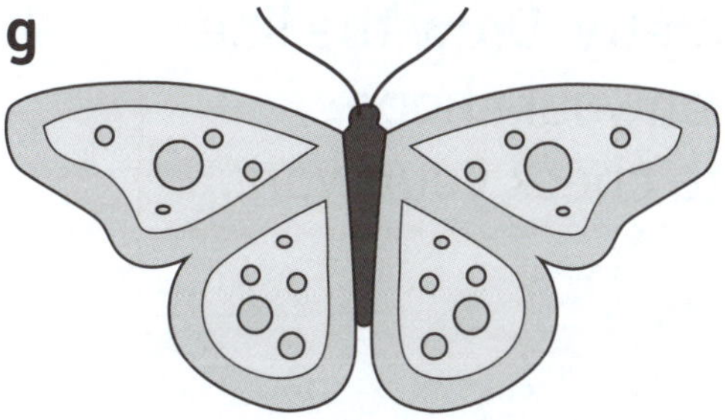

h

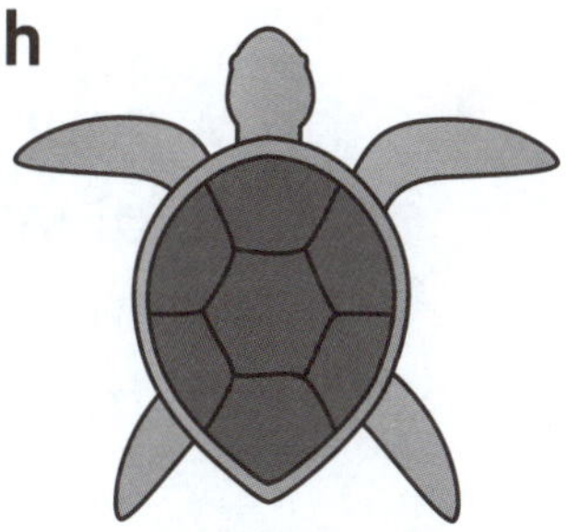

➧ *Pupil Book page 93*

Reflections

1 Complete the symmetrical shapes.

a

b

c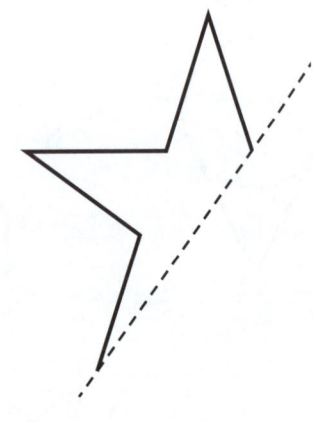

d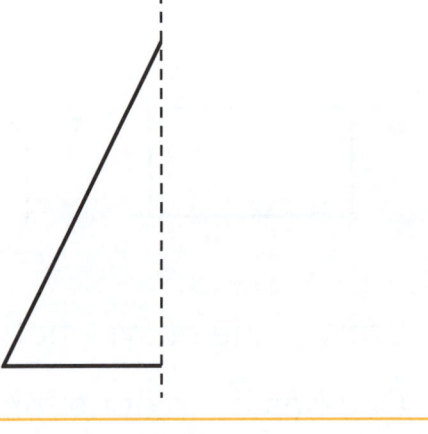

e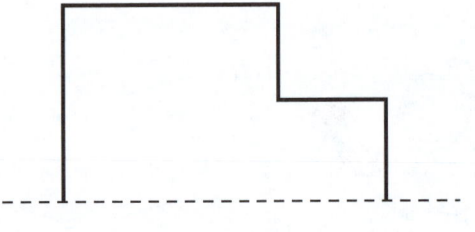

f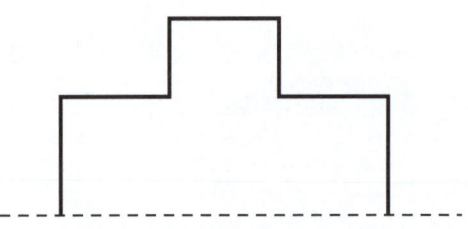

2 Draw your own shape that has a line of symmetry. Draw the line of symmetry on your shape. Use a mirror to check your work.

➡ *Pupil Book page 95*

Symmetrical patterns

1 Tick (✓) the patterns that are symmetrical.

Draw a line of symmetry on the symmetrical patterns.

2 Draw your own symmetrical patterns.

➡ *Pupil Book page 95*

Capacity and temperature

The litre

We can measure amounts of liquid in litres.

This container holds 1 litre.

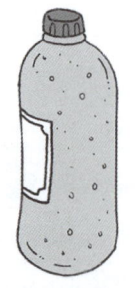

This bottle holds 2 litres.

This bucket holds 10 litres.

1 How much do you think each container holds?

Circle the correct measurement.

a	Electric kettle		about 2 litres about 20 litres
b	Large pot	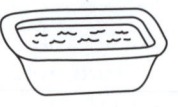	about 5 litres about 20 litres
c	Plastic baby bath		about 1 litre about 10 litres
d	Bath		about 10 litres about 100 litres
e	Rainwater tank	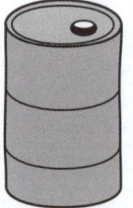	about 5 litres about 500 litres
f	Measuring jug		about $\frac{1}{2}$ litre about 10 litres

➡ *Pupil Book page 97*

Estimate and measure

1   Estimate the capacity of each container. Tick (✓) a box to show this.

Check with a litre container if possible.

Colour the box with the correct measurement.

Container	Less than 1 litre	About 1 litre	More than 1 litre
a			
b			
c			
d			
e			
f			

➡ *Pupil Book page 98*

Temperature

1 Write warm, hot, cool or cold under each picture.

a

_____

b

_____

c

_____

d

_____

e

_____

f

_____

2 Circle the item that is hotter in each pair.

a

b

3 Draw something that is colder than your hand.

➡ *Pupil Book page 99*

More about time

Tell the time

1 When do you do these activities?

Draw hands to show the time on each clock.

Draw a little ☀ or 🌙 to show if it is day or night.

The first one has been done for you.

➡ *Pupil Book page 101*

Quarter hours

1 Write the time.

The first one has been done for you.

a

quarter past 11
_____

b

_____

c

_____

d

_____

e
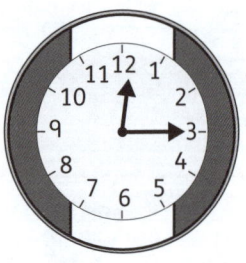
_____

f

_____

g

_____

h
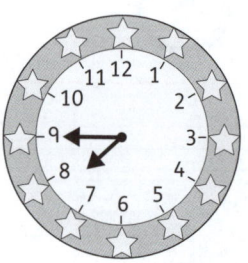
_____

➡ *Pupil Book page 102*

What is the time?

1 Write the time shown on each clock.

2 o'clock

half past 11

---

a

_____

b

_____

---

c

_____

d

_____

---

e

_____

f

_____

➡ *Pupil Book page 103*

More times

1 Draw the hands on each clock to show the time.

a

5 past 3

b

10 past 8

c

25 past 4

d

20 minutes to 12

e

5 minutes to 5

f

25 to 7

2 Write the time shown on each clock.

a

_____

b

_____

c

_____

d

_____

e

_____

f

_____

➡ *Pupil Book page 104*

Write digital times

1 Write the numbers on each clock to show the time.

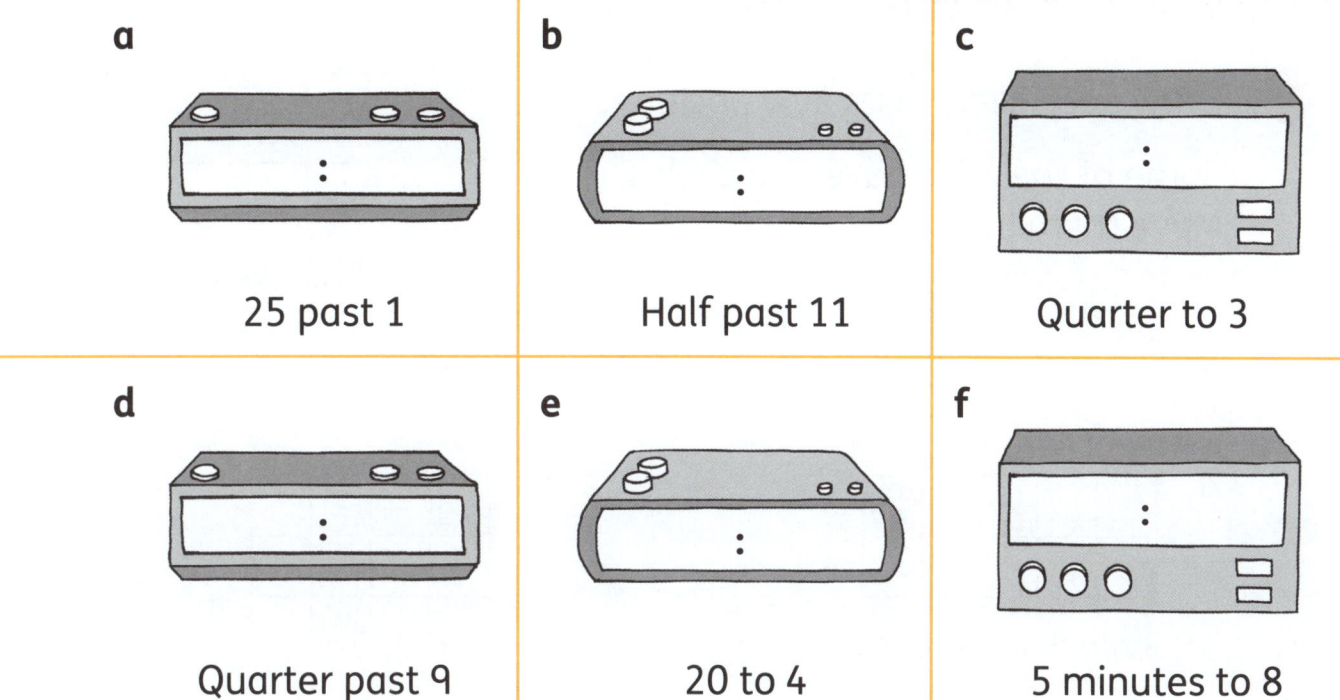

a

25 past 1

b

Half past 11

c

Quarter to 3

d

Quarter past 9

e

20 to 4

f

5 minutes to 8

2 For each pair of clocks, choose a time.

Show the same time on both clocks.

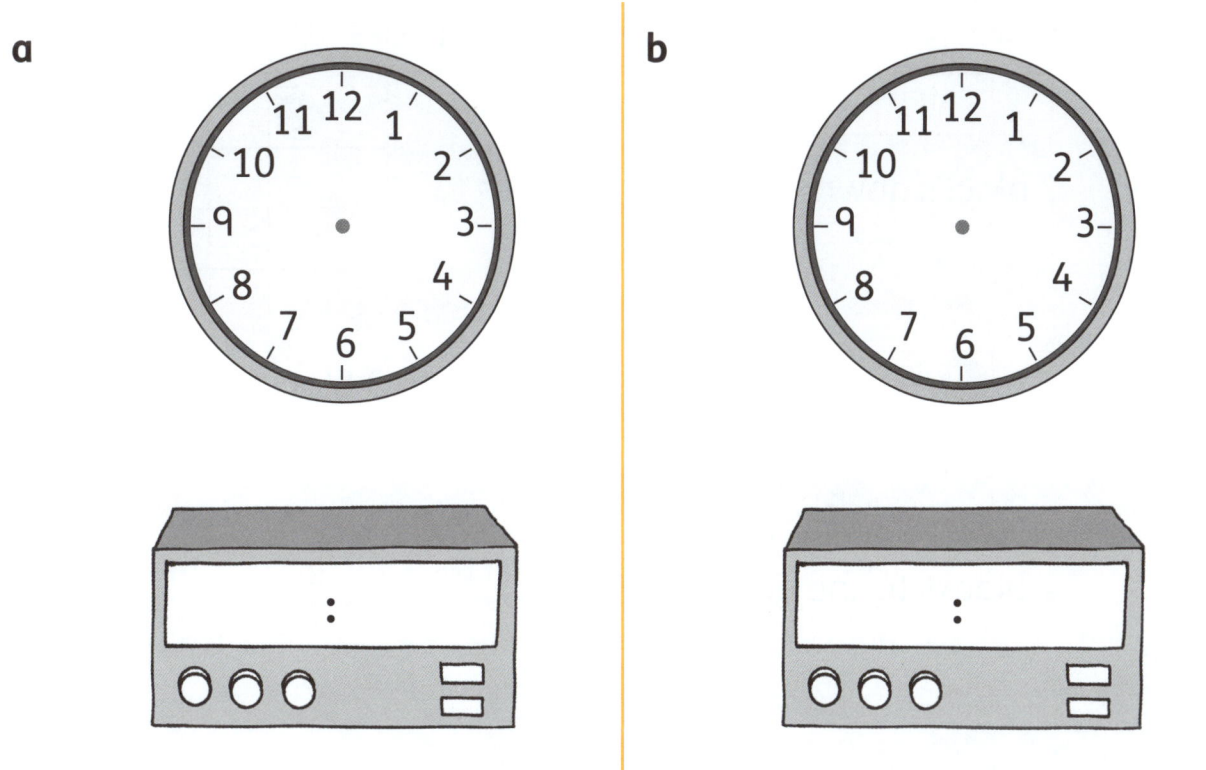

a

b

➡ *Pupil Book page 105*

UNIT 19 Position and movement

Moving in a straight line

1 Draw each circle in its new position.

Some of the grids have two circles. Move them both.

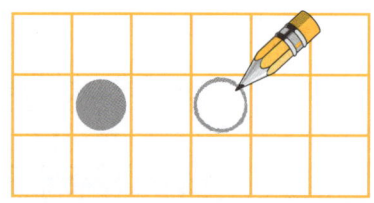

2 blocks to the right

---

a

3 blocks to the left

b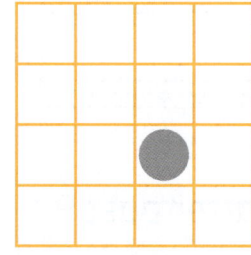

2 blocks up

c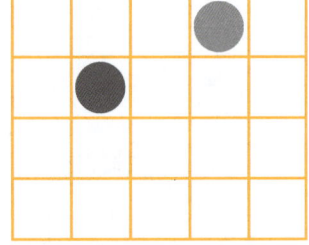

2 blocks down

d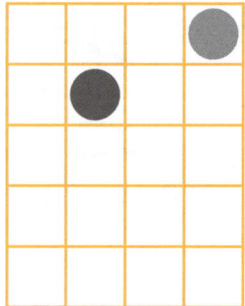

3 blocks down

e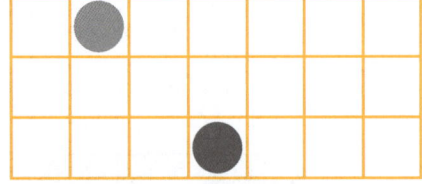

2 blocks to the right

➡ *Pupil Book page 107*

Turning left and right

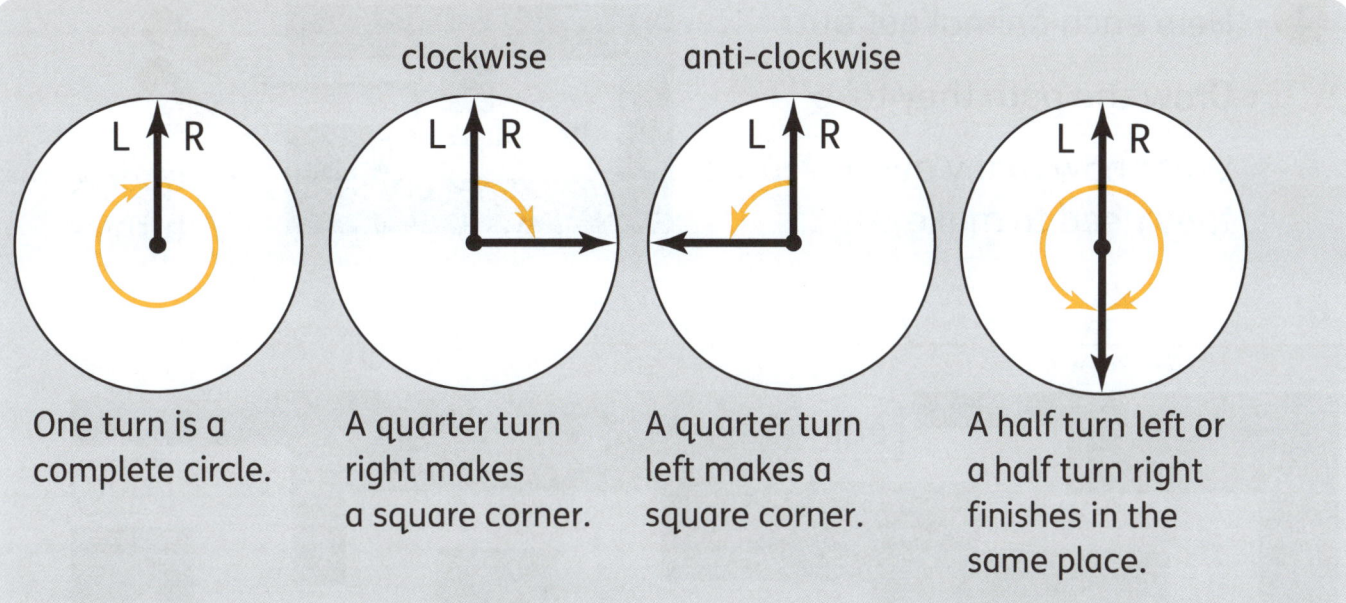

clockwise   anti-clockwise

One turn is a complete circle.

A quarter turn right makes a square corner.

A quarter turn left makes a square corner.

A half turn left or a half turn right finishes in the same place.

1 Draw where each arrow will be after a quarter turn right.

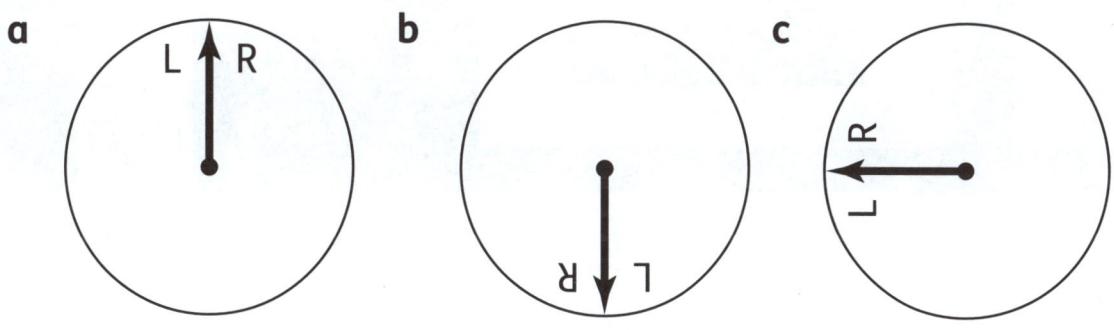

a   b   c

2 Draw where each arrow will be after a quarter turn left.

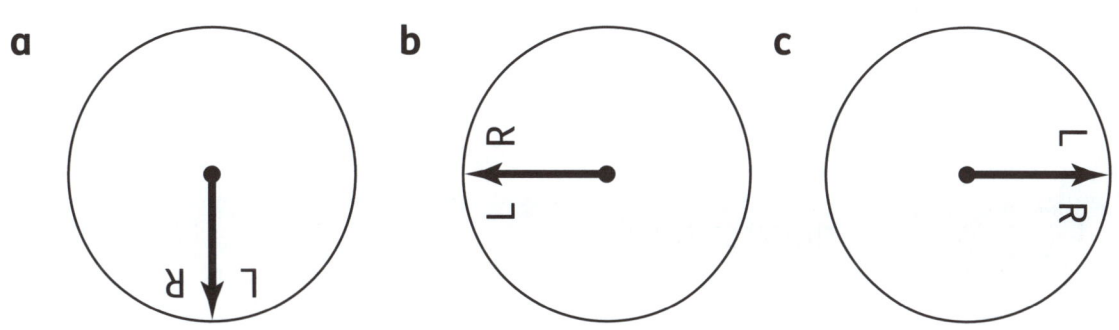

a   b   c

➡ *Pupil Book page 108*

How many turns?

1 Help each animal get out.

Draw the path they take.

Write how many quarter turns they need to make.

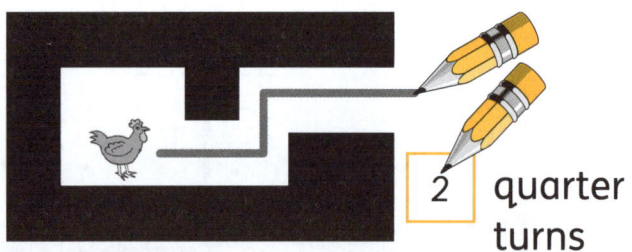

2 quarter turns

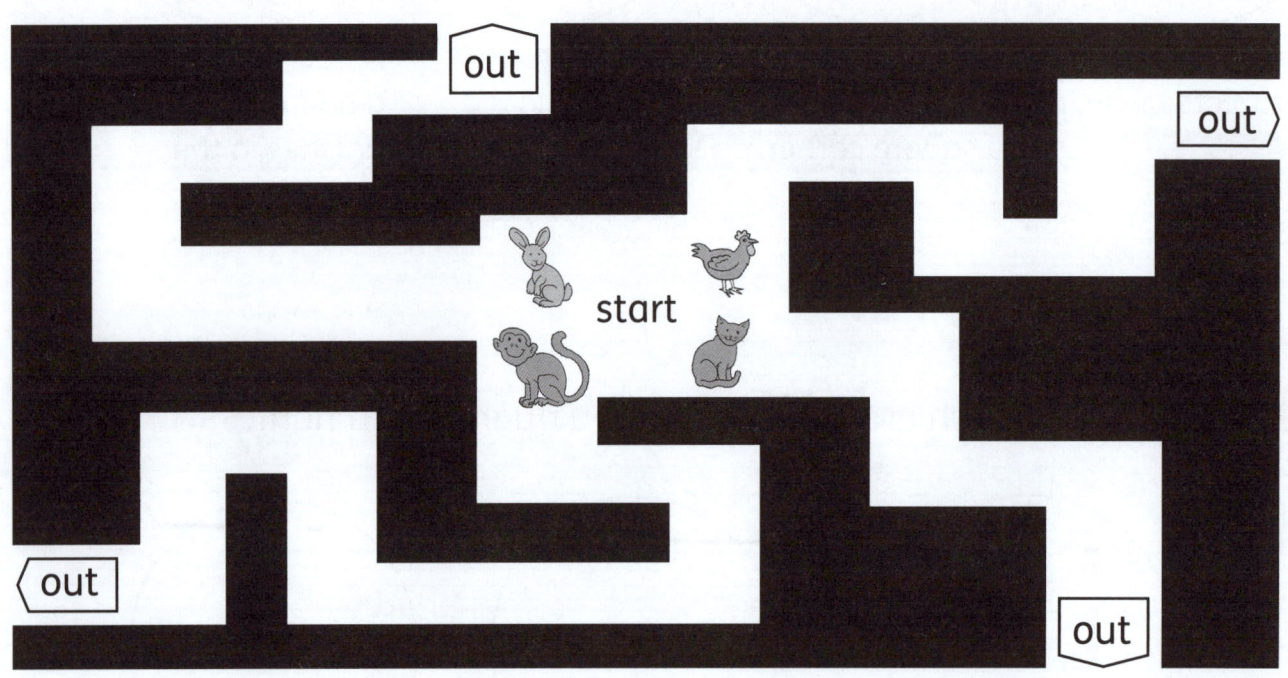

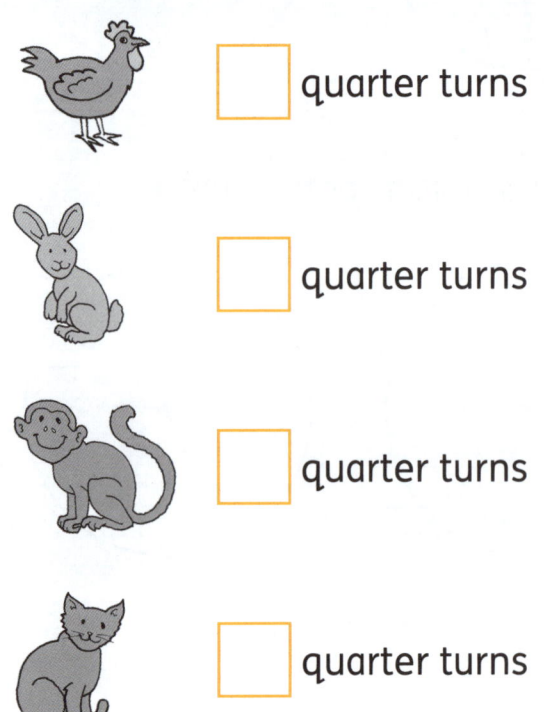

quarter turns

quarter turns

quarter turns

quarter turns

➡ *Pupil Book page 109*

Quarter turns

1 Draw the path from the start to the tree.

How many quarter turns did you make?

2 Draw the path from the house to the finish.

How many quarter turns did you make?

3 Which path has more quarter turns?

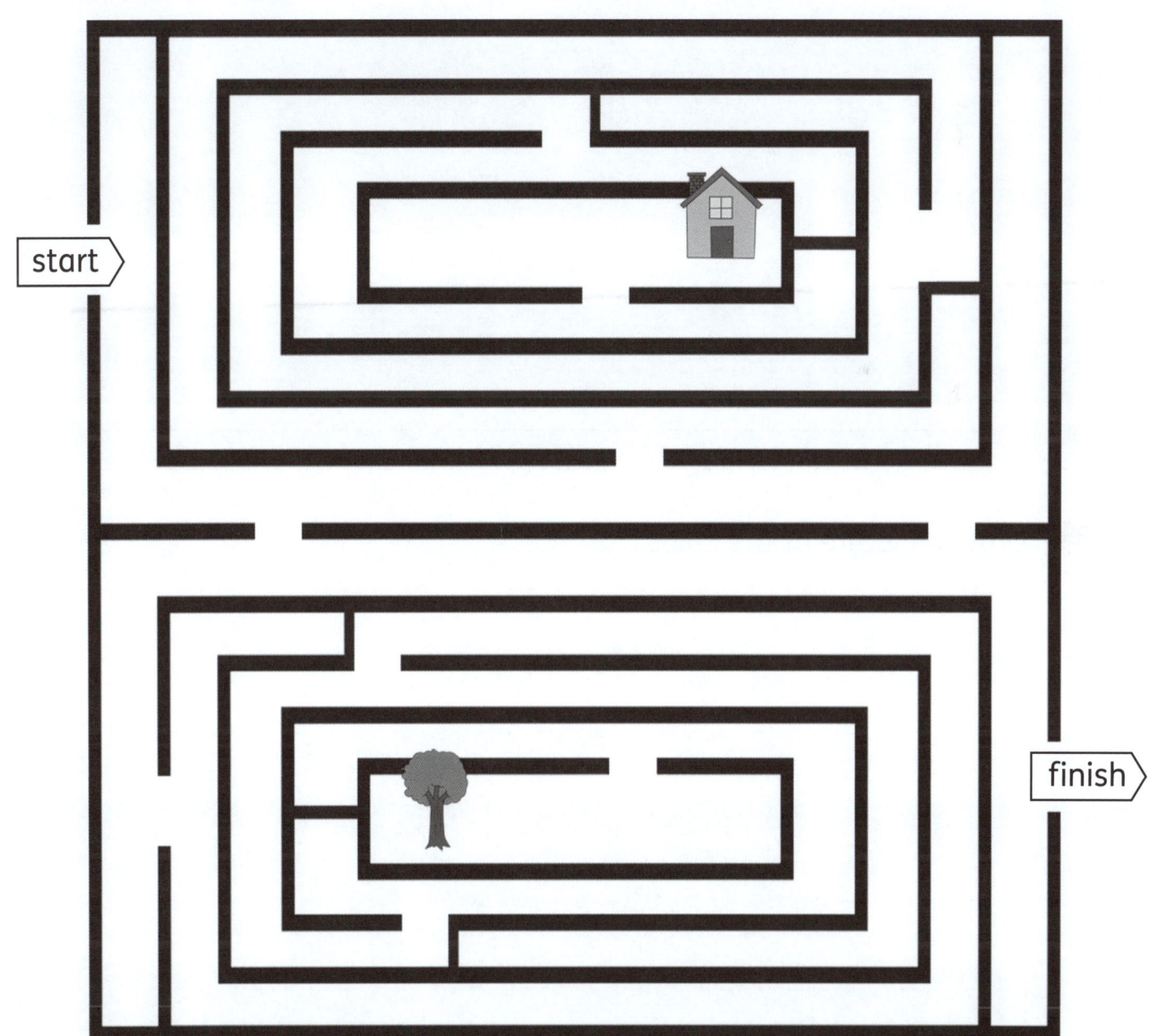

➡ *Pupil Book page 109*

Our currency

1 Complete the sentence:

In our country, the money we use is called _____

and _____.

2 Draw a picture of the notes.

3 Draw a picture of the coins.

➡ *Pupil Book page 112*

How much change?

1 Draw coins to show your change from 20 cents.

I buy ...	My change from 20 cents is ...
15 cents	
12 cents	
17 cents	
9 cents	
13 cents	

2 Work out how much change you would get.

Money given	Total to pay	Change
$10	$1	
$10	$7	
$10	$5	
$20	$12	
$20	$4	
$100	$60	

➡ *Pupil Book page 114*

How much is it worth?

Use the key to work out how much each pattern is worth.

Write the amount in cents (c).

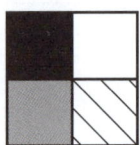

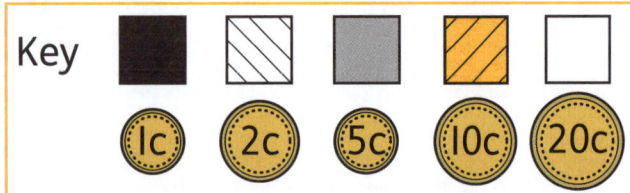

Key

| 1c | 2c | 5c | 10c | 20c |

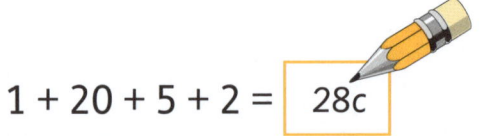

$1 + 20 + 5 + 2 = 28c$

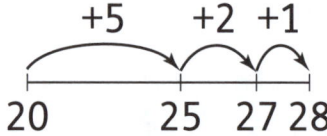

---

1 Write how much each pattern is worth.

a

☐ + ☐ + ☐ + ☐ = ☐

b
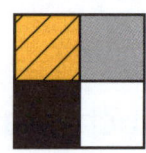

☐ + ☐ + ☐ + ☐ = ☐

c
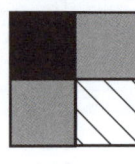

☐ + ☐ + ☐ + ☐ = ☐

d

☐ + ☐ + ☐ + ☐ = ☐

2 Make your own patterns to match each amount.

Use all the squares.

a 38 cents

b 47 cents

➡ *Pupil Book page 114*